Battleground General

El Alamein

Battleground General

El Alamein

Alistair Smith

Pen & Sword
MILITARY

First published in Great Britain in 2012 by
Pen & Sword Military
an imprint of
Pen & Sword Books Ltd
47 Church Street
Barnsley
South Yorkshire
S70 2AS

ISBN 978-1-84884-689-0

A CIP catalogue record for this book is available from the British Library

Typeset in 11pt Ehrhardt by
Mac Style, Beverley, East Yorkshire

Printed and bound in the UK by
CPI Group (UK) Ltd, Croydon, CRO 4YY

Pen & Sword Books Ltd incorporates the Imprints of Pen & Sword Aviation, Pen & Sword Family History, Pen & Sword Maritime, Pen & Sword Military, Pen & Sword Discovery, Wharncliffe Local History, Wharncliffe True Crime, Wharncliffe Transport, Pen & Sword Select, Pen & Sword Military Classics, Leo Cooper, The Praetorian Press, Remember When, Seaforth Publishing and Frontline Publishing.

For a complete list of Pen & Sword titles please contact
PEN & SWORD BOOKS LIMITED
47 Church Street, Barnsley, South Yorkshire, S70 2AS, England
E-mail: enquiries@pen-and-sword.co.uk
Website: www.pen-and-sword.co.uk

Contents

Maps

How to Play *Battleground General*

Playing or reading *Battleground General* could not be easier. For the first time, you decide how the campaign progresses. You make the choices that would have been open to the generals in the campaign. You determine the outcome of the campaign by making the right choices at the right time. You finally get the chance to try out some of your own theories on how the campaign should have been fought.

After reading the opening entry, you will be given a series of choices. Each choice takes the story to its next logical phase. Simply turn to the new **Decision** number and continue to read the narrative and description of the outcomes of your choice. Keep reading and making choices until the campaign is won or lost. You can always choose the historical timeline; *Battleground General* incorporates that into the choices offered. All of the other choices and their outcomes are speculative, based on what would probably have happened if the generals had chosen other courses of action.

You can choose either of the commanding generals, taking control of the action and determining the outcome of the campaign. Bear in mind, the book is designed so that the opposing general will try to out-think you; he will respond to your choices. Do not expect the opponent to give in easily; he will be aggressive at times, forcing you to react to his actions. When necessary, the opponent will be on the defensive and may try to extricate his forces before they are eliminated.

The storyline can either be played out to its logical conclusion, or it may end prematurely if a situation has developed where one side would naturally have decided to withdraw rather than face outright defeat and annihilation.

Introduction

In many respects the campaigns in North Africa, which would last between 1940 and 1943, had started as a sideshow. In the aftermath of the defeat of France in 1940 Axis control stretched all the way from Vichy French Morocco in the west to Italian Libya in the south. The prize asset of North Africa was Egypt and its Suez Canal. Control of the Mediterranean was the ultimate goal.

An Italian offensive launched on 13 September 1940 into Egypt from Libya had faltered within days. This precipitated the first British offensive, which lasted from 9 December 1940 until 7 February 1941. This too petered out as units were taken from North Africa to prop up Greece and ultimately Crete, both of which would be lost.

By 1941 Rommel had appeared on the scene in North Africa. With limited resources he would transform the fortunes of the Axis in the theatre. He launched his own first offensive on 24 March 1941, managing to penetrate the border regions of Egypt. Stalemate was achieved by late May. A renewed British attempt was mounted on 18 November 1941, which pushed Rommel back almost to where he had begun, deep into Libya, by the end of the year.

Rommel's last major offensive took place on 21 January 1942. This time it swept deep into Egypt, threatening Alexandria and Cairo and the vitally-important Suez Canal. In addition to this, the successes of Rommel threatened a potential link-up with Axis forces driving deep into southern Russia. The vast Middle Eastern oilfields were under threat; if they were to be lost Britain's hopes of continuing the struggle, largely alone in Europe and the Mediterranean, would be over.

Rommel knew that time was not on his side; he had to reach the Suez Canal soon, but resistance stiffened and by 7 July he had reached the outermost limits of his advance at El Alamein, short of Alexandria.

Rommel had faced a succession of Allied commanders; now he faced a new one, in the shape of Lieutenant General Bernard Montgomery. Montgomery had just ten weeks to reinforce, reorganise, re-equip and retrain his forces before launching his own offensive.

Montgomery knew that Rommel's lines of supply were stretched to breaking point. He knew they lacked aircraft and tanks and that petrol was limited. Above all, the Allies had access to the German secret codes.

Despite all of this Rommel launched an attack on Alam el Halfa on 30 August 1942. Montgomery was ready for this and counterattacked, driving Rommel back beyond El Alamein. It was the opportunity he had been waiting for; the new line offered Montgomery secure flanks. He could now prepare for his own major offensive.

Montgomery resisted Winston Churchill's increasingly desperate demands to attack. He knew he needed time to be certain of success. Churchill wanted the offensive to be launched over the full moon period in September but Montgomery chose the October full moon. He had seven weeks left to prepare Operation Lightfoot.

Carefully Montgomery concealed the preparations and the build up of men and equipment. By the middle of October his Eighth Army had swollen to over 200,000 men, supported by 1,351 tanks. Against him Rommel could muster 100,000 men and 510 tanks. Many of the infantry were Italian, as were 300 of the tanks.

There would be another distinct advantage for Montgomery in the forthcoming battle. Rommel's health was failing and from late September he was on sick leave in Germany. Facing the huge barrage and assault which began at 2140 on 23 October was General Georg Stumme. Stumme had been well briefed by Rommel and stuck to his orders. On 24 October Stumme was carrying on a reconnaissance of the front lines when he ran into British troops; although he was not hit he died from a heart attack. Temporarily General Ritter von Thoma took command of the Axis forces until Rommel could be rushed back to North Africa.

If you wish to take the role of Montgomery begin reading from page 1. If you wish to take the role of Rommel on page 111.

Montgomery, Bernard Law, Lieutenant General

Montgomery was born in 1887 in London, with his father working in Tasmania. He had an unfortunate childhood. His military career was nearly cut short at Sandhurst when he narrowly escaped expulsion after a fight with a fellow cadet.

By 1914, posted as a lieutenant in the Royal Warwickshire Regiment, he saw service in Belgium, being badly wounded but earning the DSO. In 1915, now a major, he was involved in training work, but in 1916 he took up a role as a Staff Officer. He was involved in the battles of the Somme, Arras and Passchendaele.

By the end of the war he had risen to the temporary rank of lieutenant colonel and between the wars he gradually rose in rank, reaching brigadier general in 1938, having served in India, Palestine and Egypt. He was now in command of the 3rd Infantry Division, which was deployed to Belgium in 1939. His division managed to extricate itself largely intact from Dunkirk. Montgomery's new role was defensive until the middle of 1942.

Montgomery had a long running feud with Claude Auchinleck, who by this time commanded the Eighth Army and was also CIC Middle East Command. Churchill had visited Auchinleck in August 1942 and had decided to replace him. Alexander would take over as CIC and Gatt as the commander of the Eighth Army. Gatt, however, was killed onboard an aircraft bound for Cairo and a new commander was needed.

Montgomery had just been given Alexander's old job as commander of British forces and earmarked for Operation Torch. Churchill was persuaded by the Chief of the Imperial General Staff to offer the command of the Eighth Army to Montgomery.

Montgomery officially took over on 13 August 1942; his actions over the next few weeks would transform the fortunes of the Eighth Army. He focussed on reorganisation, reinforcement and retraining. He ensured the thirty-mile front at El Alamein was reinforced in depth. He asked for two new infantry divisions and sought to coordinate the activities of the army, air force and navy.

Significantly, he anticipated where Rommel was likely to strike next. Rommel would attack the heights at Alam Halfa in an attempt to turn the left flank of the Eighth Army. Montgomery was ready for him when the attack came on 31 August. Despite Churchill ordering him to counterattack once Rommel had been defeated and forced to withdraw, Montgomery remained convinced that he needed time to launch his own offensive on his own terms.

As we know, the (second) battle of El Alamein was launched on 23 October 1942. Only violent rainstorms prevented Rommel's army from being totally destroyed after twelve days of fighting. Some thirty thousand prisoners were taken.

In time, Montgomery chased Rommel across Libya and into Tunisia. Rommel counterattacked on 6 March 1943 at Medenine, but was beaten back. After outflanking Rommel at the Mareth Line later that month, the struggle for North Africa was nearing its end. The Allies had already launched Operation Torch and a link up with Montgomery was imminent.

Attention now turned to Sicily and eventually the Italian mainland. A new rival emerged in the form of Patton, with whom Montgomery would clash throughout the remainder of the war. Montgomery left Italy in December 1943 bound for England and preparation for Operation Overlord; the Allied liberation of mainland Europe.

Montgomery would command British and Canadian forces in the east around Caen. Most of the more powerful German formations were thrown at this front, which ultimately allowed US forces to launch Operation Cobra; the breakout from Normandy.

The balance of forces in France had shifted in favour of the US, leading to Eisenhower taking over as the Grand Forces Commander (in addition to being Supreme Commander). Montgomery was incensed at this, but Churchill repaid his contribution so far by appointing him as a field marshal.

Montgomery's 21st Army Group focused on the coastal approach to Germany, battling all the way into Belgium. He was the architect of Operation Market Garden; a bold attempt to seize a series of Dutch bridges leading into the heart of Germany.

In December 1944 Montgomery was instrumental in ensuring the Allies recovered from the shock of the last German offensive in the west, in the Ardennes. Montgomery's 21st Army Group crossed the Rhine in force on

24 March 1945, a move that led to the encirclement of the Germans in the Ruhr.

On May 4 1945 Montgomery accepted the surrender of all German forces in Northern Germany, Denmark and the Netherlands.

Although Montgomery is associated with the battles in Normandy, Operation Market Garden and the surrender on Luneberg Heath, it is El Alamein that remains his crowning victory. Montgomery's plan for the battle was clear:

- To cut two corridors through the Axis minefields
- The first would run southwest towards Miteiriya Ridge and the second to the north of the ridge
- Armour would then break through the corridors and destroy the Axis armour
- Meanwhile, diversionary attacks would be launched against Ruweisat Ridge, in the centre and at the far southern end of the Axis lines.

Montgomery saw the operation as having three key phases:

- The break in (creating the corridors)
- The fight (defeating the Axis armour)
- The breaking (shattering Rommel's forces)

Montgomery expected the battle to last for twelve days. If you wish to take the role of Montgomery then turn to **Decision 1**.

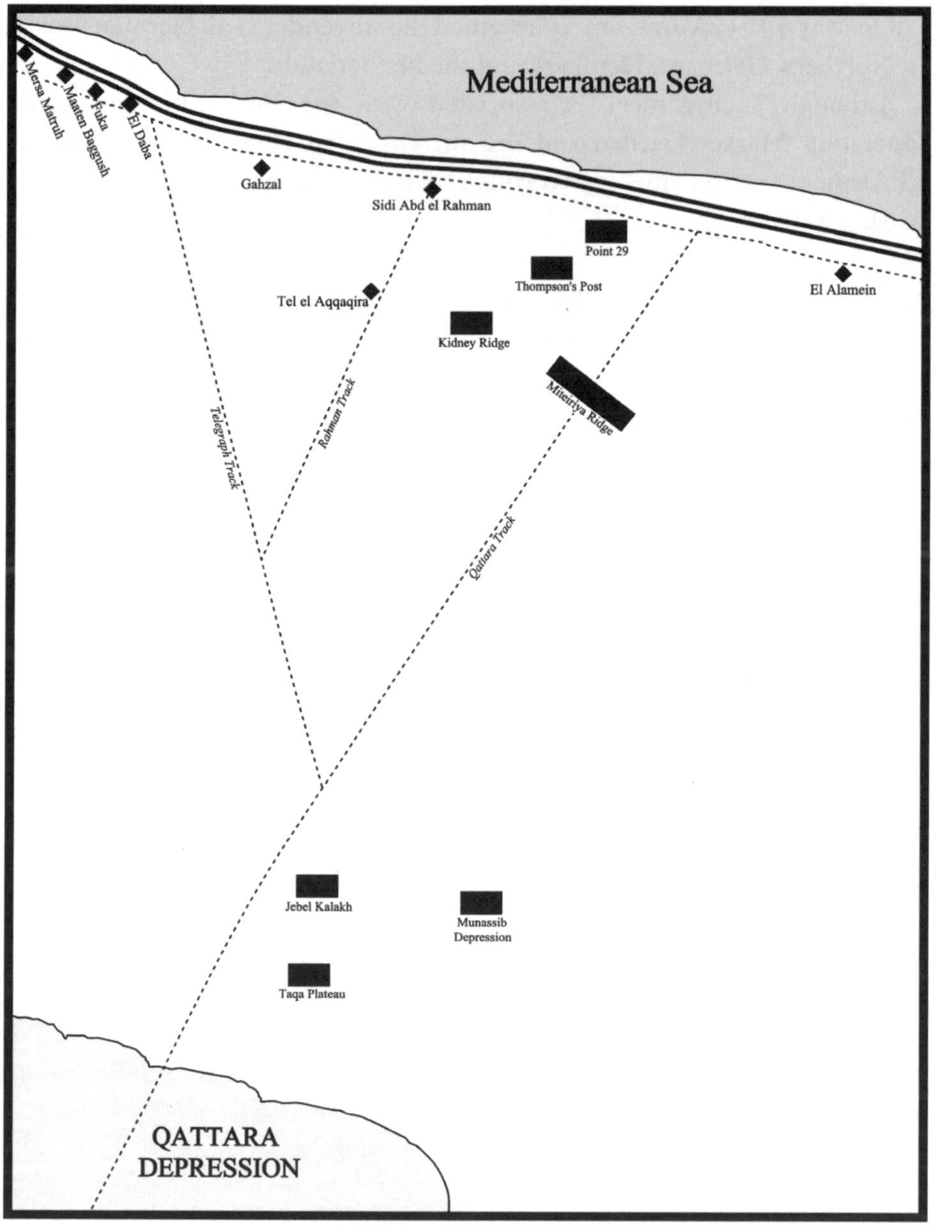
Mediterranean Sea
Mersa Matruh
Maaten Bagush
Fuka
El Daba
Gahzal
Sidi Abd el Rahman
Point 29
Thompson's Post
El Alamein
Tel el Aqqaqira
Kidney Ridge
Miteiriya Ridge
Telegraph Track
Rahman Track
Qattara Track
Jebel Kalakh
Munassib Depression
Taqa Plateau
QATTARA DEPRESSION

Battleground General: El Alamein – British

1

2140 hours 23 October 1942

It is a calm and clear evening, the sky is still bright thanks to a full moon. Operation Lightfoot, your long awaited offensive against Rommel's Afrika Korps, is about to commence.

At 2140, Egyptian summer time, 882 guns open fire along the enemy front. The bombardment continues for twenty minutes and after that the artillery focuses on specified targets ahead of your advancing infantry. You have planned your bombardment to last for five and a half hours.

Your immediate concern is to break through the vast German minefield belts, known as the Devil's Gardens; ironically most of the mines are British. Many of them are anti-tank mines, but there is a scattering of anti-personnel mines too.

Your plan is for the infantry to advance through the minefields, the engineers to follow them to clear a path, and then for the armour to punch through to the other side and engage the enemy tanks.

Your main attempt is in the northern sector and the four infantry divisions of XXX Corps get underway at 2200. You hope for a breakthrough by dawn. To the far south Horrock's XIII Corp's 7th Armoured, supported by a Free French brigade, is mounting a diversionary attack against the Italian Folgore Division. Your intention is that the diversion draws in the German armoured reserves to the south. In the centre the 50th Infantry Division is making another diversionary attack, as is the 4th Indian Infantry Division.

With growing tension you await reports. At 0200 you receive news that X Corp's armour is moving into the minefields.

By dawn XXX Corps has managed to take the majority of their objectives. The 2nd New Zealand and the 51st Highland Divisions are fully in control of Miteiriya Ridge. The 9th Australians have also crossed both minefields and to the south of XXX Corps the 1st South Africans have linked up to cover the flank of Miteiriya Ridge.

Yet despite this the armour has not yet broken out into open ground. These are desperately needed to cover the 'crumbling attacks' you have arranged for XXX Corps and to bring the enemy armour to battle. For now you need to focus on this vital northern sector. You have a number of options.

If you want to order the infantry to dig in and hold for the day, giving the armour a chance to cross the minefields in force then go to **Decision 45**. If you want to push what little armour you have through forward then go to **Decision 89**. If you want to work forward slowly, covering the mine clearing and the armour's forward movement by artillery, go to **Decision 132**.

2

With reluctance you give the commanders permission to fall back. Perhaps another attempt could be made tonight. By then, however, the element of surprise will be gone. You know that the enemy will attempt to strengthen any threatened sector of the front, but what are the options?

You had hoped that by starting at three points on the enemy front his reserves would be drawn away piecemeal and here in the north you would have sufficient concentration of forces to win a decisive victory and break Rommel's army. There is still time to reconsider your options, but this time is running out fast.

If you want to change your mind and order the 10th Armoured Division forward through the New Zealand positions then go to **Decision 231**. If you want to try and mount a renewed attempt to get your armour through then go to **Decision 174**. If you want to shift your attention to the south and hold Miteiriya Ridge for now, go to **Decision 211**.

3

'Alright. A night attack, but that does not mean that if the situation presents itself then you should attempt everything to break through. Is that clear?' you tell Lumsden.

'Absolutely,' he replies.

Shortly before 1700 you receive a report that elements of the 1st Armoured Division have broken through in the north. One brigade is believed to have reached a depression confusingly known as Kidney Ridge. This is at the northern-most edge of their intended objective. Further south the progress is disappointing and a half-hearted attack was launched at 1600 by the 8th Armoured Brigade. Now turn to **Decision 4**.

4

2100 hours 24 October 1942

The main operation to get the 10th Armoured Division beyond the Miteiriya Ridge is scheduled to begin in an hour. However Lumsden has confirmed that the 2nd Armoured Brigade is on Kidney Ridge. It has been an incredibly frustrating day. At least the success of the infantry has so far outweighed the failings of the armoured units. By your exacting schedule the armour is now a full twenty-four hours behind. Nonetheless, if the 10th Armoured Division, under Gatehouse, can get out alongside the 1st Armoured Division then the next phase of the operation can commence. This will allow the 'crumbling' of the enemy defensive lines and the driving of the enemy armour out into the open. To some extent you feel that you still have the initiative. You decide to turn in for the night and hope that the morning will bring good news. Now turn to **Decision 176**.

5

You pull on your uniform and prepare yourself for the short trip to Lumsden's HQ. You know it will be a very difficult meeting.

In the half light the roads are alive with traffic and you can clearly see gun flashes off to the west. Lumsden has obviously been warned of your imminent arrival and is waiting for you outside his command truck.

'Lumsden, what precisely is happening? I'm given to understand that the 10th Armoured has not moved forward,' you begin, staring him in the eye.

'It appears that the 10th has hung back, then they were hit by a Stuka attack that set fire to a column of supply trucks. The flames acted like a ranging marker for the enemy artillery. The 8th Armoured Brigade requested that he pull back. Meanwhile the New Zealanders, who had gone forward ahead of the tanks, were stuck. I must say that I concur with my commanders that the crews are untrained for this sort of thing,' Lumsden explains.

'Untrained for what?' you yell.

'Sitting out in the open on the forward slopes of a ridge. It is not the job of armour to take on anti-tank guns,' Lumsden answers.

'Is that right?' you reply.

If you want to order Lumsden to get Gatehouse's division forward and leave it to him then go to **Decision 177**. If you want to talk to Gatehouse yourself then go to **Decision 6**.

6

'Gatehouse, what are you doing? Why aren't your brigades forward?' you demand.

'Neither brigade has made sufficient progress through the minefield. It seems that they are both likely to be caught disorganised if we proceed. The corps commander strongly advises withdrawal to the former positions,' he replies.

'Does he? Well you cannot lead your men from the rear can you? I demand that you fight your way out. Is that clear?' you order.

'Yes sir,' he replies feebly.

'If you don't have the stomach for it then I'm sure that I can find someone that does,' you snap, ending the conversation.

Now turn to **Decision 178**.

7

Desperately unhappy with the situation, you are now in a position that seems to indicate that the whole operation has been a failure. You are not threatening the enemy's line of concentration. You cannot begin your crumbling operations and it seems that your armour is stuck. To the south, the other bad news is that the XIII Corps, under Horrocks, has fallen back. The whole campaign seems doomed. If you want to try and rectify the situation then go to **Decision 137**. If you feel it is too late to try and salvage anything then go to **Decision 179**.

8

Your intelligence reports suggest that the enemy armour is gathering in strength to the west of your positions. So far there have been some attacks by twenty to forty tanks. These are isolated attacks and seem to be somewhat uncoordinated. Your crumbling attacks along the enemy front line are beginning to work. The Trento Division to the south has been forced to give ground but on the coastal strip the Australian 9th Division, supported by the 51st Division, has made very little progress against the 164th German division or the Italian 7th Bersaglieri.

To the south there is disquieting news. The Germans have committed the 21st PZ and the Ariete Division in the area around the Munassib Depression. The 44th Division and the Free French, supported by the 7th Armoured, are having a tough time of it. You are sure that they can hold, but it seems that the Germans think they have contained your forces at the moment. If you are happy to continue to wait, go to **Decision 95**. If ou feel that something needs to be done to force the issue then go to **Decision 138**.

9

The 1st Armoured Division begins to realign itself to advance towards the north. Intelligence reports tell you that the Axis front is held by the 164th Light Division and the 7th Bersaglieri. To the west, in reserve, are the 90th Light Division and the Italian Trieste Division, although they are some way back from the front.

It is concerning you that the 15th PZ and the Littorio divisions have not yet been committed by the Germans. This move will give the Germans a great deal to worry about. If they lose the coastal strip the main roads leading west will be lost and considerable numbers of their men will be cut off. You aim to coordinate the northward move with the advance of the 9th Australians on the coast and support from the 51st Highland Division. Now turn to **Decision 97**.

10

1900 hours 25 October 1942

It takes some time for the units to pull back out of the salient and make their way to the defensive line to the west of the Miteiriya Ridge. You wonder if you have made the right decision, but at least positioning your armour here means that they can have all the support needed by your infantry and artillery. Added to this, your fuel situation is more manageable and the Germans will be forced to waste more of theirs if they choose to attack you here.

It concerns you that the impetus of your offensive has been lost to a degree and you need to decide what to do about this. If you think that you should wait and see what the Germans choose to do about the threat your forces pose then go to **Decision 143**. If you feel you need to build up a new reserve of troops for a major push then go to **Decision 184**.

11

2300 hours 25 October 1942

Reports from patrols indicate that there are open channels in the minefields around Point 29. A prisoner has been captured with a sketch map of the minefields.

At midnight lead elements of the 2/48 Battalion set off to overrun enemy outposts, whilst artillery pulverise the main enemy position. Carrier-borne Australian infantry sprint forward, but there is fierce opposition. By dawn, supported by thirty tanks from the 40th Royal Tank Regiment, Point 29 is in your hands and nearly 250 German prisoners have been taken. Now go to **Decision 55**.

12

1200 hours 26 October 1942

A counterattack by the 15th PZ Division, the 16th Light Division and elements of Italian divisions is developing around Point 29. Your men are ready to repel the attack.

Reports seem to suggest that Rommel is beginning to draw his reserves to what he feels is the focal point of the battle; the struggle around the coastal area, particularly the Kidney Ridge position.

As the enemy begins their advance, do you want to order the attack to be broken up by artillery? If so, go to **Decision 192**. If you want to deal with the armour with your tanks and anti-tank guns then go to **Decision 220**.

13

'Lumsden, swing the 10th Armoured in behind the 1st. The tank battle is to the north,' you tell him over the crackling radio set.

'It will take a little time,' he tells you. 'Perhaps an hour.'

'As quickly as possible,' you insist.

If you think that the 10th and 1st are sufficient to deal with the enemy armour in the north then go to **Decision 14**. If you feel you need to commit the 7th Armoured from the south then go to **Decision 58**.

14

0800 hours 26 October 1942

You hear the great news that the Australians have taken Sidi Abd el Rahman. The 1st Armoured Division is screening their positions. Meanwhile the 15th PZ Division and elements of the 21st PZ Division are engaged in a pitched battle with the 10th Armoured Division.

It is now obvious that Rommel has committed the bulk of his armour to the north. Only the Ariete Division remains in the south as his armoured reserve.

If you want to move your own 7th Armoured Division from the south to help your other two armoured forces then go to **Decision 110**. If you think it best to leave the 7th where it is for now then go to **Decision 152**.

15

0800 hours 26 October 1942

The Australians have been stopped short of Sidi Abd el Rahman by a determined stand by the remnants of the Italian Bersaglieri and German 90th Light Division units.

Further south the 10th have not had a quiet night; they have been attacked by elements of the 15th PZ and the newly arrived 21st PZ Division. Lumsden has shifted parts of the 1st Armoured to assist and a tank battle is still raging around the Rahman Track.

The reports are confused and it is difficult to understand precisely what is going on. If you feel you want to assess the situation for yourself then go to **Decision 193**. If you want to try to contact Lumsden and get a fuller picture then go to **Decision 221**.

16

0600 hours 26 October 1942

The minefields have proved to be thicker than anticipated. The Italian paratroopers have been putting up a stiff resistance and your probing moves have been met with heavy defensive fire.

Just before dawn a path was cleared through the minefield and the 7th Armoured began the perilous journey, only to be met by anti-tank fire from near-obsolete Italian guns. The Italians then resorted to throwing petrol bombs at your tanks.

The weight of your attack, helped by a second breach in the minefield, is too much for the Italians. In pockets they continue to fight to the last man. You have now managed to punch two half-mile gaps in the enemy line and it appears that the remnants of the Italian division are falling back beyond the Munassib Depression.

If you want to give chase then go to **Decision 25**. If you want to hold and consolidate then go to **Decision 69**.

17

1000 hours 26 October 1942

Harding's 7th Armoured Division has engaged tanks of the Italian Ariete Division around the eastern edge of the El Taqa Plateau. The Italian tanks are no match for the British armour and have been severely mauled in the encounter.

Elements of the 44th Division are now linking up with the Free French, as the Italian Pavia Division's line begins to buckle under the threat of being encircled by your armour.

It appears that the Italians are beginning to pull back. It is possible that they will try and establish a new line that hinges off Jebel Kalakh, but this will mean that the Folgore Division will also have to shift northwest to avoid being cut off.

All now hinges on your success here in the south. If you think you have sufficient forces to exploit the situation then go to **Decision 216**. If you wish to consider shifting some of your armoured divisions from the north to reinforce then go to **Decision 187**.

18

0200 hours 27 October 1942

Lead elements of the 7th Armoured Division have pushed up into the breach created by the Free French. Progress is painfully slow, with the Italians in the sector still as determined as ever and now supported by tanks of the Ariete Division. You fear that you may have lost the opportunity here.

Working through the reports, a definitive decision needs to be made. You can either commit the 7th Armoured fully, and perhaps support it with additional forces from the north, if so go to **Decision 148**, or you must call off the attacks in the south and refocus your attention on the offensive in the north and go to **Decision 188**.

19

1700 hours 26 October 1942

Your field artillery opens up first, then the anti-tank guns, and finally the tanks. Blazing enemy vehicles dot the landscape, yet they continue to advance. Impatiently you listen in on the radio network, hoping to understand the events that are unfolding. There is a variety of terrified and excited comments from the network.

'Scratch one!'

'Pot that one to the left!'

'Under heavy fire!'

'Where's my support fire?'

The din of battle is confusing, but you sense no panic in the voices, only grim determination and a relief to be part of the action at last; a chance to fight back on at least equal terms with the enemy.

'They're breaking!' you hear.

'Who? Who's breaking?' you yell back.

'The Germans, the bloody Germans are breaking!'

'The bloody Germans are breaking General,' you tease.

Now turn to **Decision 64**.

20

1800 hours 26 October 1942

With the 7th Armoured Division in the lead Horrocks begins to advance on a broad front. They come under sporadic artillery fire; clearly the enemy has not yet been able to bring up field guns in sufficient numbers to support their attack. This will be a battle of manoeuvre in the open desert, but the enemy is well versed in encounters such as this and will be a formidable foe.

Your armour comes under fire from well-placed anti-tank guns. Several are lost at extreme range; the expert German crews are exacting a heavy toll from the very beginning. Still the 7th Armoured presses on, despite the losses.

Now in range of the enemy armour, the real battle begins. Enemy Pz IIIs emerge from hidden folds in the desert, their main armament barking and the vehicles recoiling with each shot. Casualties are beginning to mount. The 4th Light Armoured Brigade is taking heavy losses; their faster but light Stuarts are easy prey for the Pz IIIs. If you want to call off the attack then go to **Decision 65**. If you want to continue then go to **Decision 108**.

21

0600 hours 27 October 1942

The crumbling operations are going well and the 7th Armoured Division has had little to do. There is no sign of the enemy ahead; you can only assume that he has fallen back some distance to the northwest. You make the snap decision to shift the 10th Armoured to the south. This will free up the 7th to probe to the west. The 10th can operate with the infantry and cover their crumbling operations. Now go to **Decision 149**.

22

2200 hours 28 October 1942

By the end of the day all of the decisions have been made. You have taken the 1st South African Division out of the line in the north. In the south you now have two armoured divisions, two infantry divisions, the Greek Brigade and the Free French. Operations are to begin at midnight.

The infantry divisions will cover the new front from Munassib Depression to the El Taqa Plateau. The two armoured divisions, supported by the Free French, will swing west around the Jebel Kalakh and strike up the road, aiming to capture the junction of the Telegraph Track, which leads to the Rahman Track and the rear of the enemy. The Greeks will protect the flank of the infantry divisions in the east.

As is your habit you retire early, having done all you can. You make sure that the forces in the north are especially aggressive tonight to stop the enemy from shifting forces south.

It is 0200 when your chief of staff wakes you up.

'Rommel is counterattacking with the 15th PZ to the north of Jebel Kalakh,' he tells you.

If you want to go back to sleep and leave it to the commanders in the field then go to **Decision 66**. If you want to intervene then go to **Decision 109**.

23

2000 hours 26 October 1942

It takes some hours for the 1st Armoured Division to fall back without getting itself cut off. The Australian units have taken heavy casualties in the fighting withdrawal.

This timid approach and fear of German armour has made you something of a laughing stock. You have even heard someone whisper 'There goes Bolting Bernard, the scourge of reverse gear'.

You feel that no one understands how important a victory is and that you should not take unnecessary chances. If you feel you have lost control of the situation and need to tell Churchill that a victory is not yet assured then go to **Decision 234**. If you feel that there is much to be done and that you should call a conference to decide on your next actions then go to **Decision 159**.

24

0800 hours 27 October 1942

It takes some time for you to reorganise your forces. So far your armour has been somewhat disappointing and their performance lacklustre. You know that the battle should be fought in the north; it is the pivotal part of the battlefield. You have spent the last day pouring over reconnaissance reports and situation updates. Your next moves will undoubtedly decide the outcome of the battle. Above all, you need to retake the initiative, added to which you need to draw out the enemy armour and destroy it. Having done this, you can then focus on mopping up Rommel's less mobile forces. Once a breakthrough has been achieved his army is doomed. Now go to **Decision 219**.

25

1200 hours 26 October 1942

With mounting impatience you wait for news from Horrocks. He should be in clear and open ground. Reports suggest that Horrocks has moved to the west of his previous positions. It also appears that he has run into the German 21st PZ Division. Now go to **Decision 20**.

26

0600 hours 28 October 1942

You decide to pull the troops back to their start positions and rethink your strategy on the northern front.

Perhaps you need to consider pulling troops out of the front line and creating a reserve that you can launch against the enemy with a fresh offensive? Your new reserve will consist of the New Zealand Division, along with the 10th and 7th Armoured divisions. If the Australians can make some progress in the northern part of the battlefield, close to the coast, you could then unleash this reserve with the objective of destroying the remnants of the enemy armour. Now go to **Decision 155**.

27

0600 hours 29 October 1942

You order the 26th Australian Infantry Brigade back to its start positions. You determine to try again the following night, but you realise that the enemy will be ready and waiting for you. Nonetheless, Rommel is likely to think that this will continue to be the focus of your offensive. By attracting his reserves to this part of the front at least it gives you the opportunity to continue to destroy his armour piecemeal, waste his fuel supplies and give you time to build up your own reserves. Now go to **Decision 114**.

28

0800 hours 31 October 1942

Rommel's attention is clearly drawn to the coastal sector, which suits your purposes. You now need to make a definitive decision about where your attack will be made. You codename your new operation - Supercharge.

Nominally you determine that it will take place tonight. That should give you sufficient time to bring your reserves together for the operation. You then need to determine where the axe needs to fall on the enemy front.

If you want to proceed with plans to launch the offensive tonight then go to **Decision 115**. If you feel that you want to postpone for another twenty four hours, to make sure that everything is in position, then go to **Decision 157**.

29

2000 hours 1 November 1942

Your aircraft have been attacking enemy ground targets all day. Half an hour ago your artillery began to saturate the enemy front lines and reserve areas with shells. You have amassed 360 guns.

Your two main assault brigades, the 151st Durham Brigade from the 50th Division and the 152nd (Seaforth and Camerons) Brigade from the 51st Division, are due to get underway at 0100. They will be supported by the 8th and 50th Royal Tank Regiments. Once a breakthrough has been achieved the 1st Armoured Division will cut through the gap and engage any armoured reserves that Rommel has been able to muster.

The whole assault, codenamed Operation Lightfoot, depends on clearing paths through the minefield to break through to the Rahman Track, which is only 1800m behind the enemy lines. This plan must work. The 9th Armoured Brigade is detailed to make the first breakthrough ahead of the 1st Armoured Division, but although it started its approach at 2000 it is not yet in position at 0500. If you want to call off the attack then go to **Decision 223**. If you want to postpone for half an hour until they are ready then go to **Decision 30**.

30

0615 2 November 1942

The three regiments of the 9th Armoured Brigade finally get underway. It is barely thirty minutes before dawn. Their job is to attack what is literally a wall of anti-tank guns and burst through.

As your tanks close on the enemy positions the rising sun silhouettes them against the horizon. The crack of 88mm anti-tank guns is joined by the quieter thud of Pak 38s and Italian 47mm guns. A handful of the German tanks have worked their way in amongst your advancing units and are causing casualties you can ill afford.

The 9th Armoured Brigade presses on into the very face of the enemy, with losses mounting. If you think you have made a terrible mistake and it will end in defeat then go to **Decision 120**. If you want to press on, fearless of the losses, then go to **Decision 162**.

31

0800 hours 27 October 1942

The situations reports confirm that the offensive in the north has continued to make huge advances overnight. The 15th and 21st PZ Divisions are now desperately trying to hold on to ground around the Rahman Track, keeping a corridor open for the retreating infantry. On the coastal road a series of small-scale counterattacks by the 90th Light Division have kept your Australians alert.

Further south the Italian infantry divisions holding the line have fallen back nearly five miles. In the far south Horrocks has sent out probing reconnaissance missions and has scooped up several thousand prisoners that were desperately trying to retreat northwest. It seems that the whole of the enemy front in the centre and the south is in chaos. Only in the north is there any really organised and determined resistance.

Your commanders are urging an all out offensive. If you agree with them then go to **Decision 75**. If you think the battle is still to be won or lost in the north then go to **Decision 118**.

32

0800 hours 28 October 1942

Intercepts report that Hitler has given Rommel permission to withdraw from the battlefield. By now the bulk of his infantry has successfully traversed the desert tracks and the men are making their way west, by whatever means available. Some fuel has been delivered to the enemy, but not nearly enough, as abandoned trucks and tanks are evident across the whole of the desert.

Slowly but surely the 90th Light Division and the handful of remaining tanks begin to withdraw. Your armoured commanders are desperate to pursue and finally you give them permission. But it is largely too late; Rommel continues to fall back. Each time you give chase he deftly turns, stings your forward units then retreats once more. He is still a formidable foe. You can only pursue with the utmost vigour and clear Libya of the enemy, perhaps chase him into Tunisia, but only then can the desert war be truly over.

33

0800 hours 27 October 1942

Your aircraft have been in operation since first light, roaming up and down the Rahman and Telegraph Tracks at will. They have reported huge columns of men and vehicles heading northwest. Each time a sortie has been launched the enemy has fired back with scattered anti-aircraft batteries, but it does appear that the bulk of the enemy army in the south and centre is retreating to the west. You send Horrocks back to the south and order a full scale offensive along the entire front. Now is the time to finish this once and for all. Now go to **Decision 75**.

34

0800 hours 2 November 1942

The 1st Armoured Division has started to deploy, taking the remnants of the 9th Armoured Brigade under its wing. The 1st Armoured Division's 2nd Armoured Brigade attacks to the north of the salient and the 8th Armoured Brigade to the south. The 9th Armoured Brigade is in the centre, with the bulk of the 2nd New Zealand Division in support.

Rommel is characteristically adept at creating new battle groups and defensive positions. You are dismayed by the lack of progress by the 1st Armoured Division and feel that once again you may be losing the initiative. This is brought home to you when you learn that the remnants of the 15th and 21st PZ Divisions, supported by the Italian Littorio Armoured Division, have moved up to face your armour. They appear to be organising to deliver a counterattack.

If you want to order your tanks forward to engage the enemy then go to **Decision 121**. If you want them to dig in amongst an anti-tank screen then go to **Decision 225**.

35

1300 hours 2 November 1942

British, German and Italian tanks are intermingled in the confusion under the desert sun. The tank crews stalk one another, looking for that precision knock-out shot. The crews know that escape from the battlefield if their tank fails them will be a perilous state of affairs. The likelihood of survival in the carnage and destruction is low.

You determine to view what you can of the battle, moving up to the front, within the salient. Standing on top of one of the abandoned tanks of the 9th Armoured, you can just make out amongst the swirling dust clouds the flash of weapon fire. Occasionally you can hear a detonation as one more tank succumbs to enemy fire. It is all over so quickly and soon elements of the armoured divisions are heading back to refuel and rearm. They are jubilant and claim huge losses for the Germans and Italians.

In fact your losses have been about equal; in all you have lost seventy or so vehicles, virtually the same as Rommel. You can afford these losses though and he cannot. You can now focus on enlarging the gap in the enemy line and creating the opportunity for the 7th Armoured to burst through into open country. Now go to **Decision 163**.

36

2030 hours 2 November 1942

You have four brigades ready to be used as a reserve for the next thrust. They will work in cooperation with XXX Corps. The brigades, the 5th Indian, the 151st, 154th and 5th New Zealand, will attempt to enlarge the salient by attacking key German and Italian positions to the south and north of the salient. At the same time the Australians in the north will renew their operations around a position called Thompson's Post.

The new offensive operations cannot possibly start tonight; the ground has not been properly reconnoitred. You will have to be content with holding your salient during the day and then launching your attack after all preparations have been made. Now go to **Decision 164**.

37

0200 hours 4 November 1942

The 5th Indian Brigade reaches its objective, four miles to the south of Tel el Aqqaqir, and they report that the enemy line is abandoned. The retreat has all the appearances of being a hasty one. Equipment and supplies have been left behind and they have only encountered very limited opposition from the retreating enemy units.

This is good news. Perhaps Rommel has finally decided to abandon the battlefield.

As the 154th Brigade begins to move forward, shortly after 0600 hours, the Argyll and Sutherland Highlanders march unopposed into Tel el Aqqaqir, only encountering sporadic shelling from a retreating enemy.

As dawn arrives an early reconnaissance flight suggests that armour is moving northwest, away from the Rahman Track and towards the railway on the coast. You order a second reconnaissance mission to try and establish the identity of the unit. You suspect that it is what remains of the 21st PZ Division. Now go to **Decision 165**.

38

0800 hours 4 November 1942

Elements of the 1st Armoured Division have spent the day pushing the remnants of the 21st PZ Division west and they have achieved a penetration of another eight miles. The 7th Armoured has been in tough fighting with the Ariete Armoured Division. By early morning the Italian formation has virtually ceased to exist. Elsewhere both the Littorio Division and the Trieste Motorised Division have been wiped out.

As far as you can tell the Bologna and Trento Divisions are trying to fight their way out of the trap, but they are not motorised units. Rommel has clearly pulled out his mobile German units and left the Italians to their fate.

By late morning it is obvious that you have won a significant victory. Pursuit is now the key. You have at least a twelve-mile gap to pour troops through. You determine to send the 1st Armoured to El Daba along the coast, the 7th Armoured to Galal, and the New Zealanders towards Fuka. Now go to **Decision 83**.

39

1900 hours 4 November 1942

Your focus on the Ariete Division has allowed considerable numbers of German troops, tanks and vehicles to flee to the west. Nonetheless you determine to extricate your armour and order them to pursue the enemy.

The 1st Armoured Division will head for El Daba down the coastline. The 7th Armoured's objective is Galal, along the railway line. The New Zealanders are ordered to make their way across the desert, towards Fuka.

Your aircraft report that there are no significant lines of enemy defences ahead, so speed is essential before Rommel can prepare positions to hold off your probing attacks. The New Zealanders are due to reach Fuka by the morning of 5 November. You hear that they have come under bombardment from units attached to the remains of the 15th PZ Division and what is more they have also run into another minefield ahead.

If you want to order them to halt then go to **Division 126**. If you want them to press on then go to **Division 167**.

40

0800 hours 5 November 1942

The 1st Armoured Division ran out of fuel at dawn this morning. They are still sixteen miles short of Bir Khalda on the road to Mersa Matruh. The 7th Armoured is still short of Sidi Haneish.

Meanwhile, the New Zealanders and the 8th Armoured Brigade of the 10th Armoured Division are still moving towards Fuka. It is going to take some time before the fuel trucks catch up with the pursuing divisions. Now go to **Decision 238**.

41

1600 hours 5 November 1942

You order the New Zealanders to enter the minefield. Your hope is that it is a dummy minefield and you have not sent the men to their deaths. Nervously you await news, busying yourself with reports from the other units that are working their way through the desert to the coast. You hope that they can reach the coastal towns before the rearguard of Rommel's forces; the more that you can catch and destroy now, the less Rommel will have to work with when he does eventually turn and fight you again.

After what seems like an eternity you hear from the New Zealanders that it was in fact a dummy minefield and that they are safely through it. There are no signs of enemy resistance and they are forging their way towards Fuka. Now go to **Decision 169**.

42

It is a comprehensive if not a crushing victory. Rommel's main force of German troops has managed to slip away to the west. It has been an entirely different story for the Italians; ultimately, thousands of them have been taken captive.

Over the next months you will pursue your quarry 1500 miles across the North African coast. You determine to advance slowly and deliberately, knowing that Rommel is still capable of turning and defeating you.

If only you had known that Rommel was down to his last dozen tanks. Tobruk finally falls for the last time on 1 November 1942. At Gazala Rommel turns and fights, but only briefly before heading west again. Benghazi falls eleven days later. Rommel holds at El Aghella, but now it is too late for him. The Anglo-American landings in French North Africa mean he must fall back once again.

Finally, on 23 January 1943, your troops enter Tripoli and the last action of the desert war is fought at Zuara, against the rearguard of the 90th Light Division. On 12 February 1943, two years after the arrival of Rommel in North Africa, the remnants of his army slip across the Tunisian border.

43

1400 hours 6 November 1942

Light armoured units attached to the New Zealanders have made incredible gains over the course of the morning. The 10th Armoured, crashing through attempts to hold them up along the coastal road, have reached Sidi Barrani to find several hundred Italians waiting to surrender. They had been left behind in Rommel's desperate attempt to escape your trap.

Meanwhile the 7th Armoured has swung south to take Fort Capuzzo and Sidi Azeiz. By the end of the day the whole Egyptian border is cleared of enemy troops. You know that the pursuit is coming to an end. You can only risk sending armoured cars and highly mobile units, since the supply situation becomes more problematic as you head west.

The victory is complete. Rommel's army has been shattered. You are particularly pleased that so few German units have escaped, but you know Rommel will try to reform his army and sooner or later he will turn and fight you one more time.

44

1200 hours 6 November 1942

Incredible news has reached you at headquarters and you decide to risk flying west to Mersa Matruh to personally witness the situation.

As you arrive at the dirt airfield, surrounding by cheering troops, you are whisked west to a small desert outpost. Parked under the shade of some palm trees is the unmistakable shape of a British command truck. You recognise it instantly as Rommel's own headquarters. Confidently you stride towards the vehicle; it is under armed guard and the occupants are ushered out to meet you.

The clutch of senior German and Italian commanders step out first and then Rommel himself. The rapid movement of your armoured division cut him off overnight. His personal bodyguard and supply trucks lost him in the darkness and a British armoured car troop found Rommel here just three hours ago.

Your victory is now complete. Rommel has been captured, his army destroyed and the liberation of North Africa is considerably closer than it has ever been. You look forward to the opportunity of debriefing Rommel in person. Such a formidable foe deserves respect, but you are relieved that you will not have to face him on the battlefield ever again.

45

0550 24 October 1942

The New Zealanders on Miteiriya Ridge are firmly entrenched but they lack anti-tank guns and tank support. The minefield is not yet clear, so it is impossible to get vehicles through to them. The Australian Division has reached its objectives, but strong enemy resistance in the 51st Division sector is holding up the creation of the northern path for the armour through the minefields. To the south of the New Zealanders the South Africans have linked up, except on their right, where they are still 500m from their objectives. The Eighth Armoured Brigade, part of the 10th Armoured Division, is on Miteiriya Ridge, having got through the southern corridor. They are coming under anti-tank fire and the commander wants to withdraw behind the Miteiriya Ridge.

If you want to order him to push forward, given that the northern corridor is not yet cleared, then go to **Decision 173**. If you want to give him permission to fall back then go to **Decision 210**.

46

1225 hours 24 October 1942

It maddens you that the armour has been so inactive and the leadership so lacklustre. You can only be proud of your infantry so far, but the armoured punch has been a huge disappointment. Even now the armoured commanders are counselling that the attacks should take place tonight and not this afternoon. You are terribly aware that every lost minute gives the enemy breathing space to reorganise. What will become of your armour then, should they breakthrough into the face of a determined and well-prepared enemy defensive anti-tank line? Patiently you read all of the signals and reports. If you want to order Lumsden, commanding X Corps, to get the 10th Armoured out by this afternoon then go to **Decision 232**. If you decide to agree that a night attack is better, then go to **Decision 3**.

47

1720 hours 24 October 1942

News has reached you that in the north elements of the 1st Armoured Division have broken through the front and lead units have reached the depression known as Kidney Ridge. Contradictory reports suggest that the brigade is in fact far short of the objectives. In the south of the Miteiriya Ridge the 8th Armoured Brigade has launched a reconnaissance in force across the ridge, but apparently very little progress has been made. Still the armour is failing to push forward with sufficient spirit and vigour. You wonder what you can do about this. You manage to contact Lumsden at X Corps HQ.

'Why so little progress Lumsden?' you demand.

'We are moving forward. I expect positive news soon,' he tells you.

'Convey this to the commanders. Unless I see some significant progress then they will be replaced. Your position is by no means safe either. Do I make myself clear?'

'Very clear,' he tells you.

Now turn to **Decision 4**.

48

2100 hours October 24 1942

With nothing to work on you fear the worst. Only your infantry has acquitted itself well today. The armour, so vital to the operation, has failed you.

'The Prime Minister for you,' your staff officer announces.

'Yes Prime Minister,' you reply after clearing your throat.

You have to admit to him that the operation is only successful up to a point so far. You have no idea of the progress of your armour.

Churchill is livid and he tells you that he has moved heaven and earth to provide you with the resources for victory. Why is it that he cannot find a commander to give him the victory? You try to explain, but he is clearly not in the mood to hear excuses.

If you want to try and persuade him that he will have his much-needed victory then go to **Decision 213**. If you are unsure that you can do this then go to **Decision 234**.

49

You convene the conference for 0330. Leese, commander of the infantry XXX Corps and Lumsden, commanding the armoured X Corps, are both in attendance, along with de Guingand, your chief of staff.

'My New Zealanders are beginning to move southwest from the Miteiriya Ridge, but its success depends on the 10th Armoured Division breaking through. One of the regiments of the 8th Armoured Brigade is already into the open,' Leese begins.

'However, Gatehouse wants to pull the regiment back onto the ridge. The 24th Armoured Brigade is not yet through the minefield or the gap, but hope to be by dawn,' Lumsden adds.

'So you're suggesting that you want to withdraw both of these brigades back behind the minefields and give up everything that has been gained?' you ask.

'Gatehouse believes that his situation out in the open will be very unpleasant and he might suffer high casualties. I have to say that I agree with him,' Lumsden continues.

'And where is Gatehouse?' you demand.

'His HQ was eleven miles from the front,' Lumsden admits.

If you feel the opportunities have been lost and you agree with Lumsden then go to **Decision 50**. If you disagree with him and want to talk to Gatehouse then go to **Decision 93**.

50

'Very well Lumsden, we must try and renew our efforts. Do what you can to reorganise your commands,' you tell him.

You sit back in a chair and ponder your decision. So much has been gambled on this offensive and for it to fizzle out now, so soon after it started, is a terrible blow. Some of your commanders are so out of touch with their brigades. The confusion and incompetence of some of the old Eighth Army's operations seem to have returned, despite all of your efforts.

You know that by the morning the initiative will have passed to the enemy. If you want to stick to your decision then go to **Decision 7**. If you want to change your mind and get Gatehouse on the telephone then go to **Decision 6**.

51

0700 hours 25 October 1942

It seems that your threats and cajoling have paid off. Confirmed reports suggest that both the 1st Armoured and the 10th Armoured are now fully deployed ahead of the Miteiriya Ridge and that the last elements of German defensive positions have been cleared from Kidney Ridge.

The crumbling operations can now begin. You eagerly await news that the Germans have committed their armoured resources in the north. Your two armoured divisions can expect to be attacked by the 15th PZ and the Littorio Division at any time.

The Germans must be well aware of the fact that your northern assault is your main effort. What worries you is a concerted assault by both of the enemy divisions; hopefully they will attack piecemeal. The 1st and 10th Armoured are now sitting out front, with the German 164th in the north and the Italian Trento Division to the south. Somewhere ahead of them are the two Axis armoured formations.

If you want to order the 1st and 10th to sit tight and wait for the enemy to react then go to **Decision 8**. If you want to order a reconnaissance in force to the west then go to **Decision 52**.

52

Reconnaissance units soon begin to run into elements of 15th PZ. Reports suggest that the Germans are trying to mass their armour for a determined attack on your two armoured divisions. Gatehouse contacts you direct, unable to get Lumsden, who he believes is visiting the Kidney Ridge area to inspect the new defensive line.

'What do you want me to do?' Gatehouse asks you.

You wonder whether the situation is right for a major battle with the enemy armour.

If you want to order Gatehouse to pull back to the defensive line and let the Germans mount their attacks on your new positions then go to **Decision 95**. If you want to tell Gatehouse to bring up the rest of his armoured division and order Lumsden to support him with the 1st Armoured then go to **Decision 96**.

53

The Greek Brigade is a relatively strong formation, with three battalions of infantry supported by a field artillery regiment, machine guns and some engineers. You hope that their pressure will aid your efforts in the south. A breakthrough here will pull in the German and Italian armoured reserves and prevent the enemy from bringing them up to the north to counter your main effort.

You need to make sure that the attacks in the south are carried out with sufficient vigour to put the enemy off balance and make him think that the danger lies to the south.

Facing you in the south, at the extreme end of the enemy line, is the Pavia Division, with three infantry regiments. Covering their northern flank is the Italian Folgore Division, with three regiments of paratroopers, supported by an artillery regiment. To their north is the considerably tougher German Paratroop Division, under Hermann-Bernhard Ramcke.

Where do you want the main effort to be made? If you want it to be against the Pavia Division then go to **Decision 140**. If you want it to be against the Folgore Division then go to **Decision 182**.

54

1600 hours 25 October 1942

Despite Lumsden's fears that the enemy will hit his exposed flank if he moves to the north, he has made reasonable progress by the late afternoon. The 10th Armoured Division, on the flank of the advance, has reported that increasing numbers of enemy tanks and transports are massing to the west. A spotter plane confirms that they are cutting across the desert from the west.

There is better news from the north; the German 164th Division has given up a good deal of ground in the east, as they are pressed by the Australian Division. Clearly they are concerned about being cut off if your 1st Armoured Division breaks through to the coast.

By dusk Lumsden reports that his lead tanks are no more than five miles from Sidi Abd el Rahman. Just to the south of the town is the railway and just north of the town is the coast itself.

If you want to order Lumsden to continue to press on then go to **Decision 215**. If you feel he has accomplished enough today and want the Australians to press on overnight from the east then go to **Decision 15**.

55

It is clear that the actions over the past couple of days have been ruinous as far as the enemy is concerned. The attrition on their front line units indicate that the Italian Trento Division is down to about fifty-percent strength. The German 164th has already lost the equivalent of two battalions. In addition to this both the 15th PZ and the Littorio have lost numerous tanks, albeit piecemeal so far.

You glean from intercepted communications that the Axis forces are short on rations and that fuel will only last for another three days. Yet how can it be that they are still resisting you?

Reports also seem to suggest that the enemy will try and focus on pushing you out of the Kidney Ridge area, which threatens their lateral supply route, the Rahman Track.

If you want to dig in around this position with tanks supported by infantry then go to **Decision 12**. If you feel that the time is right to push west and cut the track then go to **Decision 180**.

56

2100 hours 25 October 1942

A tremendous artillery barrage erupts and plasters the enemy front lines, support trucks and rear areas. Briggs' 1st Armoured Division is first to get off, just behind the creeping barrage. It is led by the 2nd Armoured Brigade; a formidable force of nearly 160 tanks.

Half an hour later the Australians begin moving forward. At first the enemy response is limited, but as the Australians advance they begin to come under intense fire.

Did you manage to take Point 29? If so, go to **Decision 99**. If you failed to take Point 29 then go to **Decision 141**.

57

0800 hours 26 October 1942

Clearly the enemy is far more resilient than you had feared. The offensive has stalled and in places the enemy is beginning to push your lead elements back.

The 1st Armoured Division is under grave threat now; unconfirmed sources suggest that the enemy has brought up the 21st PZ Division from the south to stem your attack.

You need to act quickly otherwise all of the gains you have made will be lost. Worse still, the 1st Armoured Division could be ruined as a fighting force. If you wish to pull back then go to **Decision 23**. If you want to throw in the 10th Armoured Division then go to **Decision 67**.

58

0800 hours 26 October 1942

The die is cast as you commit the 7th Armoured Division to the fight in the north. On the coastal road the enemy is falling back beyond Sidi Abd el Rahman, which is now fully held by the Australians and elements of the 1st Armoured Division. The 10th and 7th have been in joint operations against enemy armour close to the Rahman Track.

Along the rest of the front it seems that the enemy is beginning to fall back. Rommel is concerned that his slower moving, un-motorised units will be cut off if they cannot reach the coast road before you block their path. Victory it seems is close. Now go to **Decision 68**.

59

1500 hours 26 October 1942

The battle in the north is one of attrition; so far neither side has shown signs of breaking or being able to land a blow that is really telling. Reports now seem to suggest that the enemy has finally committed his remaining armoured unit to the north. Your aircraft have been attacking ground targets on the Rahman Track all morning. They are believed to be the bulk of the 21st PZ Division and elements of the Ariete, who are moving up from the south.

It is clear that Rommel sees the north as the crucial area of the battlefield. You decide that you need to re-establish a reserve that you can use, either to plug gaps or to use them as a cutting edge should the opportunity arise.

You pull the New Zealand Division out of the line, along with the 10th Armoured. You also make the decision to shift the 7th Armoured from the south. Now go to **Decision 153**.

60

0600 26 October 1942

By dawn the Free French are precariously holding on to a salient in the Italian defensive line. Accurate enemy artillery fire has driven back the bulk of the 44th Infantry Division; they are too exposed in a minefield and are taking heavy casualties. The French report that they have seen Italian tanks manoeuvring near the El Taqa Plateau. As yet the 7th Armoured Division has not moved up to support the French. If you want to order them to do so then go to **Decision 17**. If you think it wiser to wait for darkness again before committing the armour then go to **Decision 61**.

2

If posting from outside of the UK please affix stamp here

Pen & Sword Books
FREEPOST SF5
47 Church Street
BARNSLEY
South Yorkshire
S70 2BR

61

2000 hours 26 October 1942

Pressure has been mounting on the Free French all day, despite throwing your air assets at troop concentrations forming up to attack them. The Italians have been working hard to re-establish their line and show little signs of panic. Increasing numbers of Italian tanks have been seen; they have not yet launched a major attack on the French, but are softening them up with long-range fire.

During the day Horrocks has tried to reinforce the French with additional infantry and has been using his artillery to break up enemy threats against the French before they can develop.

The 7th Armoured is ready to be unleashed, but still you hesitate, wondering if the enemy has had too much time and has been able to move the 21st PZ Division down as a major mobile reserve. Now go to **Decision 18**.

62

1400 hours 26 October 1942

In the baking heat of the desert the 7th Armoured Division begins to advance. At first it runs into small groups of retreating Italian infantry. These are soon scooped up and sent east by your infantry. As the advance continues the tanks encounter well-placed anti-tank gun positions. The defensive line being set up by the enemy seems to be further south than you expected. Nonetheless the enemy defensive line is still weak and Horrock's men punch through with minimal losses.

Ahead they can see several dozen enemy tanks spurting to take up positions, and lines of trucks are visible, frantically refuelling the armour. The German 21st PZ Division, led by von Randow, is only a shadow of its former self, yet it cannot be underestimated. The tank crews are veterans of many engagements.

Harding, the 7th's commander, contacts Horrocks, who in turn contacts you. 'Do we engage?' is the simple question.

If they should engage then go to **Decision 217.** If they should pull back and take up defensive positions then go to **Decision 105.**

63

2200 hours 26 October 1942

The 10th Armoured Division is finally in place and ready for offensive action. There have been numerous breakdowns and the Springbok Track is littered with vehicles being hastily repaired to be despatched south. Nonetheless, the bulk of the force is ready and backed up behind the 7th Armoured.

As the armour begins to move the 7th Armoured sweeps onto an abandoned El Taqa Plateau; there is only sporadic firing as trailing Italian units return fire.

The 10th Armoured have swept along the rear of the Pavia Division's defensive line and it appears to have been largely abandoned.

The 44th Infantry Division, under covering fire from the artillery, has broken through the minefield and is occupying the positions. The Greek Brigade reports that the Folgore Division has also pulled back.

Meanwhile the Free French have taken up defensive positions on the El Taqa Plateau, freeing up the 7th Armoured to continue their advance.

If you want the force to consolidate their positions then go to **Decision 107.** If you feel you want them to press on and find the new enemy defensive line then go to **Decision 149.**

64

2000 hours 26 October 1942

The enemy armour and infantry are in full retreat. They have pulled back out of artillery range and appear to be heading north and northwest. Nothing now stands between your forces and the flank and rear of the enemy line.

To the west is a track that leads past Jebel Kalkh and into the heart of the enemy rear. Horrocks can now either strike west and pick up the track, if so go to **Decision 235**, or he can remain in position and guard the infantry as they crumble the enemy defensive positions to the north, go to **Decision 21.**

65

0600 hours 27 October 1942

You pull back the 7th Armoured Division from its forward positions, covering them with concentrated artillery fire. You take the decision to reinforce them with the 10th Armoured Division. You give clear instructions to Horrocks. If he feels that he can initiate an attack as soon as the 10th Armoured arrives then he is to do so. Now turn to **Decision 149**.

66

0600 hours 28 October 1942

You wake up to a shambles. The 7th Armoured Division has been shattered and the 10th Armoured is falling back towards the El Taqa Plateau, having lost at least half of its tanks. The Italian Littorio Division, along with remnants of the 21st PZ and the Ariete Division, has overrun the Greek Brigade and have effectively split your army in half. It is an absolute disaster. You feel that you need to break the news to Churchill. Now go to **Decision 234**.

67

1400 hours 26 October 1942

The 10th Armoured Division pushes forward to reinforce the 1st Armoured.

'Get across to every man that if every tank hits and disables an enemy one then we must win. If we can destroy each of those tanks before Rommel can tow them away then our win becomes all the easier,' you tell your commanders.

You know that great determination is essential. You need to keep the initiative and maintain the pressure on Rommel. You need to proceed relentlessly and never give the enemy a chance to reorganise. Your initial reports suggest a chaotic situation, as your tanks are fed into the battles against the enemy armour. You know that this kind of battle not only means a great loss of tanks, men and material, but also depletes your relatively limited stores of ammunition. At this rate of expenditure the ammunition is likely to last for ten days, but you anticipate that Rommel will not survive another seven. Now go to **Decision 198**.

68

1900 hours 26 October 1942

The remnants of the German 164th and the Italian Bersaglieri have been reduced to a series of small pockets in the north. What remains of Rommel's armour has been mounting ever more desperate counterattacks against your lead units. The Italian Littorio Division has virtually ceased to exist, as it seems to have run out of fuel; abandoned tanks and vehicles litter the desert. The 7th Armoured has swung to the south to try and cut off the retreating enemy. If you want to continue to fight solely in the north then go to **Decision 31**. If you want to order an all out offensive then go to **Decision 75**.

69

1200 hours 26 October 1942

Horrocks is convinced that if he strikes to the west he can reach the Rahman Track. He begs you to attach another armoured division to his southern attack force. If you want to agree with him and send the 10th Armoured south to help him then go to **Decision 20**. If you feel that the fight should focus on the north and that he should pull back from his advanced positions and set up a new defensive line for now then go to **Decision 188**.

70

0600 hours 28 October 1942

The Rifle Brigade's anti-tank guns manage to hold off concerted attempts to overrun the position. By midday over thirty enemy tanks are burning in the desert sun. The casualties have been heavy and only one or two anti-tank guns are still operational. You pound the surrounding area with artillery fire and your aircraft prowl for targets.

The German and Italian armour pulls back; beaten but not yet broken. You determine to shift the focus of the battle to the north and direct the Australians to launch an assault around the Point 29 area and aim for the coast. Now go to **Decision 113**.

71

2400 hours 30 October 1942

Rommel has brought up the 90th Light Division to the coastal area. There are also signs that the 21st PZ Division is trying to form up a battle group to make a determined attack on your positions around Thompson's Post.

By early morning the 21st PZ Division has been able to launch four successive attacks, but has made very little ground. The last attack has managed to make contact with the trapped 125th PZ Grenadier Regiment. Rommel is clearly trying to dislodge you with these united counterattacks. If you want to respond with your own counterattack then go to **Decision 156**. If you would prefer to wait and see the brutal struggle out, keeping your reserves ready for your own offensive then go to **Decision 28.**

72

0830 hours 31 October 1942

'We've still got over 800 tanks left and his fuel situation is perilous. By keeping his attention here on the coast my reserves can strike him hard and decisively. One last thrust and Rommel will be beaten. Have patience, victory is assured.'

'I sincerely hope so,' Churchill begins, 'when will you launch the attack?'

'Hopefully tonight,' you tell him

'No more delays! I need a victory,' he concludes.

You still need to determine exactly where the attack will take place. The reserves are still building up and will be ready soon. If you want to make a decision and go ahead tonight then turn to **Decision 115**. If you think it prudent to postpone for twenty-four hours then go to **Decision 157**.

73

2000 hours 1 November 1942

The initial thrust of your attack is to be made by the 2nd New Zealand Division. You reinforce the division with the 151st (Durham) Brigade from the 50th Division and the 152nd (Seaforth and Cameron) from the 51st Division, along with the 133rd Royal Sussex Lorried Infantry Brigade. The division will also have the 9th Armoured Brigade under their command.

You plan to advance and clear a path through the minefield with the infantry before sending out a regiment of tanks on the left and right to break open a gap to the Rahman Track. The 1st Armoured Division will then follow through to exploit and bring Rommel's armour to battle. Now go to **Decision 197**.

74

1900 hours 26 October 1942

As darkness begins to fall the enemy armour begins to fall back; the battlefield is littered with knocked out or immobilised vehicles. Your men are jubilant. The enemy has dashed itself against your defensive lines for little or no gain. Your own casualties are relatively light. There will need to be a period of reorganisation and rethinking, which is likely to take at least twenty-four hours. It is time that you can begin to think about renewed offensive actions. Now go to **Decision 219**.

75

0900 hours 27 October 1942

With the enemy desperately hanging on in the north, still able to mount limited counterattacks, but clearly in retreat in the centre and the south, you order a full scale offensive.

You set the Rahman Track as the objective for your infantry divisions in the north, ordering the two armoured divisions to strike out as far west as possible and engage any remaining German or Italian armour.

In the south Horrocks is to attack northwest, cut the supply tracks to the rear of the retreating enemy and hold there to allow the infantry to advance and reduce any pockets of resistance. You hope there is no way back for Rommel now. He must try and salvage what he can of his army at the expense of losing the bulk of his infantry. Either that or he risks losing what remains of his mobile troops and suffering absolute defeat. Now turn to **Decision 160**.

76

2100 hours 27 October 1942

Your tanks swiftly deal with the last German armour, shielding the 90th Light Division. This remaining German mobile force expertly uses its vehicles and anti-tank guns to hold your armour off for as long as possible. It is no one-sided affair. They knock out several dozen of your tanks before sheer weight of numbers overwhelms them and they begin to surrender, just in small batches at first and then in greater numbers.

Your tanks bypass the positions, leaving the prisoners for the Lorried Infantry that is following up. They begin snapping at the heels of the tail of the retreating enemy columns.

Rommel is wily and unpredictable; each time he seems to conjure something and inflict a stinging counterattack. You call off the pursuit; your tanks are too exposed and overextended.

After a few days of reorganisation you send your men west in pursuit once again. Rommel's force has escaped with far more intact than he could have hoped for. It will give him heart to reorganise, turn and fight. You must defeat him again before the desert war is finally over.

77

0900 hours 27 October 1942

The 7th Armoured is only too eager to oblige. Attacking on a broad front, supported by Lorried Infantry units, the armoured brigades close on the 90th Light Division and the handful of tanks scraped together to form their defensive line. The attack takes a little time to develop; each passing hour allows more and more of Rommel's army to escape to the west. You wait patiently for news, hoping that the 7th Armoured will attack with sufficient élan to break the last resistance in the north and sweep you to victory over Rommel. Now go to **Decision 76**.

78

0800 hours 2 November 1942

You order the 1st Armoured Division to hold for now, sending the remnants of the 9th Armoured Brigade forward to try to enlarge the gap. Meanwhile your artillery pounds the enemy front line along the Rahman Track.

The results are slow in coming. The enemy is well dug in and seems determined to weather the bombardment. The key objectives on and around the Rahman Track are the main focus of your artillery fire. You are determined to break up any concentrations of enemy resistance and to prevent Rommel from establishing yet another line of anti-tank gun defences.

News comes through that enemy armour is moving up through the edges of the barrage. It seems that Rommel is planning to risk his tanks in a counterattack against the salient.

If you think that the artillery and your own anti-tank screen is sufficient to beat him off then go to **Decision 236**. If you believe it is now time to unleash the 1st Armoured Division then go to **Decision 121**.

79

1130 hours 2 November 1942

'They've got our range!' you hear being yelled over the radio, 'we're pulling back!'

Furiously you pound the table. Perhaps you misjudged Rommel and his determination to survive against all odds?

'Where are those tanks!' you yell.

'Which tanks? Theirs or ours?' asks your chief of staff.

'Ours of course!' you shout back.

'Unclear, probably to the rear,' he replies.

'Get them in there now! I want this attack stopped dead in its tracks!'

Now turn to **Decision 201**.

80

0800 hours 3 November 1942

The 7th Armoured Division has managed to break through the thin crust of anti-tank guns in two places. Part of the division is chasing the remnants of the 21st PZ Division northwest, in a running battle.

Reports suggest that the German armour is falling back rapidly, with two or three tanks being sacrificed as a defence against your advancing armour every two or three miles.

The other part of the 7th Division has run into stiffer opposition in the form of the Ariete Division, to the south of Tel el Aqqaqir. Little progress has been made and the Italians are fighting hard and refusing to budge.

If you want to support the 7th Armoured with infantry then go to **Decision 203**. If you want to send the 2nd and 8th Armoured Brigades to help then go to **Decision 38**.

81

2200 hours 3 November 1942

You order the 151st to hold their positions. They have at least reached the Rahman Track and their presence should mean that the enemy will have to respond to their threat. It does open up the gap to an extent and if the enemy wants to reinforce Tel el Aqqaqir, or use the track, then the 151st will have to be dislodged. Now go to **Decision 37**.

82

0800 hours 4 November 1942

The bulk of the 1st Armoured Division, including the 2nd Armoured Brigade, has chased the 21st PZ Division across the desert and has succeeded in virtually annihilating them.

The 7th Armoured Division, free to swing northwest, has trapped the remnants of the 90th Light Division against the coast and the 9th Australian Division has closed the circle. There can be no hope for the Germans on the coastal strip; it is only a matter of time before they are forced to surrender.

Your infantry and remaining armour have been pounding the Italian units, particularly the Ariete Division, for the past twenty-four hours. The combat has also sucked in the Littorio and Trieste Divisions. Your commanders report that they have effectively ceased to exist as fighting formations. Elsewhere the Italian infantry divisions, notably the Trento and Bologna, have been trying to fight their way west without water, food or transports. Your lead units have taken many hundreds of prisoners. Now go to **Decision 83**.

83

0800 hours 5 November 1942

El Daba has fallen to the 1st Armoured Division. Unfortunately the town has been abandoned by the Germans and the rearguard has slipped to the west.

The 10th Armoured Division is making slow progress and lorry-mounted New Zealanders, en route to Fuka, have come across a minefield. They report a wire fence in the desert to be blocking their way. They also believe that beyond the minefield are German positions.

If you feel that the minefield is somehow familiar then go to **Decision 206**. If you think it better to send a reconnaissance aircraft to check the situation then go to **Decision 168**.

84

0200 hours 5 November 1942

The New Zealanders have dug in around Fuka and have been holding off determined attacks by the 15th PZ Division to break through to the west. Freyburg, the New Zealand commander, reports that increasing numbers of the enemy tanks and vehicles are being abandoned, presumably due to the shortage of fuel. He is confident that the New Zealanders can hold Fuka indefinitely, but is concerned that the Germans will try and slip around his flank to the south and escape to the west.

If you think he can handle the situation on his own then go to **Decision 129**. If you want to send the 10th Armoured along the coastal road to assist him then go to **Decision 170**.

85

1300 hours 5 November 1942

You are unsure as to whether the minefield is real or a dummy. Given the fact that the enemy has been in control of this area for some time it is perfectly possible that they have constructed a minefield over the dummy field to try and fool you into entering it. Alternatively, they may not have had sufficient time to do this and the area is perfectly safe. If you want to order the New Zealanders to cross the minefield then go to **Decision 41**. If you think it is safer for them to work their way around the minefield then go to **Decision 128**.

86

2200 hours 5 November 1942

Your pursuit of Rommel is progressing well. Elements of the 15th and 21st PZ Divisions have surrendered to your lead elements. Each time, after a short fight, it became clear that the Germans were virtually out of fuel and ammunition. Losing these units are a huge blow to Rommel, who obviously had hoped to extricate them from the traps you have been attempting to set. With three armoured divisions converging on Mersa Matruh you order the divisions forward, throwing all your efforts into ensuring that the divisions remain refuelled. All other units are ordered to follow if and when possible. It will be some days before they can move, being out of position and overwhelmed with the difficulties of dealing with so many prisoners. Now go to **Decision 44**.

87

0800 hours 7 November 1942

The 10th Armoured Division entered Mersa Matruh in the early hours of the morning. Most of their infantry have been left behind along the coastal road to mop up enemy positions that were bypassed by the armour during the night.

Aerial reconnaissance reports that the Germans are digging in around Sidi Barrani, some eighty miles to the west of Mersa Matruh. You realise that this is meant to be a holding force to allow the bulk of Rommel's army to traverse the narrow passes at Sollum and Halfya.

If you want to call off the pursuit now then go to **Decision 42**. If you want the 10th Armoured to try and force Sidi Barrani then go to **Decision 172**.

88

1900 hours 5 November 1942

The 10th Armoured finally runs out of fuel some ten miles beyond Sidi Barrani. Behind them, in the wake of their sudden surge are the shattered remains of the German forces in North Africa.

Very little has managed to escape through the passes at Halfya and Sollum. The infantry attached to the 10th are engaged in sweeping up hundreds of dispirited German soldiers. Lead elements of the 7th Armoured have also halted short of the passes, feeling that their low fuel levels would jeopardise any gains likely to be made. The wily Desert Fox, Rommel, has managed to elude you, but with so little left you hope that the war in North Africa will soon be over.

89

It seems that the northern corridor is still being covered by the enemy in the 51st Infantry Division's sector. This is preventing the 1st Armoured Division from pushing forward beyond the infantry's objectives. The southern corridor is more promising. Not only have the New Zealanders taken all their objectives, but lead elements of the 10th Armoured, notably the 8th Armoured Brigade, are moving up through Miteiriya Ridge and could break out. Now turn to **Decision 173**.

90

'It is essential that we clear the last enemy blocking positions in the north,' you tell the commander of the 51st Division. 'When can you launch a determined assault?'

'I'm trying to organise things now. It is considerably harder than a simple mopping up exercise. They are well dug in and waiting for us,' comes the reply.

'What do you need?' you ask him.

'Artillery support and aircraft,' he tells you.

'I want the 1st Armoured Division out in front of your positions as soon as possible. Take every necessary step required to achieve this, is that clear?'

'Perfectly,' he confirms.

It worries you that the armour is so reluctant to get forward, even in the southern corridor. The armour should be out and covering the infantry. You can only wait for the situation to develop and for your men to prepare for another attack. Now turn to **Decision 47**.

91

1700 hours 24 October 1942

Eagerly you await news of progress. The minutes pass slowly by, with no word from the front. Your instructions and meticulous planning could not have been more clear and complete. How can it be that your armoured units are so reluctant to risk exposing themselves? Your infantry has done what could be expected of them, but the whole operation is now in doubt.

Perhaps you needed more time? Perhaps you should have replaced the commanders with someone who had more to prove and was willing to take risks? Each passing moment builds on the fear of failure. If you want to continue to wait then go to **Decision 48**. If you wish to contact Lumsden at X Corps HQ then go to **Decision 47**.

92

'Lumsden, what precisely is happening with your 1st Armoured Division?'

'I am satisfied that the 2nd Armoured Brigade is on Kidney Ridge,' he tells you.

'Very well,' you reply. 'You must continue to press forward in the south at the earliest opportunity.'

It has been a frustrating day, but the infantry have performed well after all, despite the lack of dash and spirit of your armoured forces. If Gatehouse's

10th Armoured Division can get out and deploy alongside the 1st Armoured then at least the operation can proceed as planned, despite being behind schedule. You are relieved to some extent that Lumsden is taking command of the situation and you decide to try and grab a few hours sleep. Now turn to **Decision 176**.

93

'Gatehouse, I am determined that the armour will get out of the minefield area and into the open. You are not to waver now!' you order him.

'I appreciate that,' he tells you.

'If you feel you cannot do this then I will replace you immediately, is that clear?' you continue.

'Perfectly clear. I will push the tanks forward. Just give me a little time,' he asks.

'I want 8th Armoured Brigade to push at least one regiment forward to protect the flank of 9th Armoured Brigade. Meanwhile, I want the gaps in the minefields improved so that additional forces can come up,' you order.

'It'll be done,' he agrees.

Now turn to **Decision 51**.

94

1200 hours 25 October 1942

Throughout the morning the Axis forces launch a series of counterattacks, spearheaded by the 15th PZ Division and the Littorio Divisions. It is clear that they are probing for a weakness. It is now becoming obvious that the main fight will develop around the Kidney Ridge area of the front. The Kidney Ridge area is now being firmly held by the 51st Highland Division, supported by armour. To the north the Australians are due to make an attempt on Point 29. This is an elevated artillery observation post, to the west of Tel el Eisa. If this can be taken it will put considerable pressure on the German 164th Division in the coastal sector of the enemy front. The attack is due to go in at midnight tonight.

If you want to attempt to take the position with the Australian 26th Brigade, which has been detailed with the operation, then go to **Decision**

214. If you want to order artillery and tank support to help take the position then go to **Decision 11**.

95

Gatehouse's 10th Armoured and Briggs' 1st Armoured are in good defensive positions. You nervously await news, believing that the Germans will attack. Perhaps you are wrong, as reports suggest that the 21st PZ and the Ariete are still in action to the south. It seems that the Germans are unwilling to risk attacking you with only half of their armour. Either way, the situation suits you for the time being. Your infantry are continuing to widen the breach in the enemy defensive line in the north. Your only concern is that the major armoured struggle that you anticipated has not yet taken place. If you are still happy to let the operation continue at its own pace then go to **Decision 139**. If you want to force the situation then go to **Decision 138**.

96

Lumsden promises to push his divisions forward to engage the enemy armour. If you can force the Germans to commit their armour in the north then there is every chance that they will also be forced to move their armoured reserves in the south to help stop any breakthrough.

As time passes you still only hear that scattered and uncoordinated attempts to stop your attack are being made. As yet the Germans have not concentrated in force. You feel that you need to force the situation one way or another. Lumsden is able to confirm that the bulk of his armour is out and ready, having taken up advanced positions.

If you want to get the 1st Armoured Division to swing around to the north and coordinate attacks on the coastal strip with the two infantry divisions facing the enemy in that sector then go to **Decision 9**. If you want Lumsden to push his armour west and find the enemy tank concentrations then go to **Decision 180**.

97

0600 hours 26 October 1942

Finally, it appears that your long anticipated Axis counterattack is beginning to materialise. There are strong indications that the enemy is building up elements of the 15th PZ Division, the 164th Division and supporting Italian units.

The flank of the 1st Armoured Division is not yet under threat, but the weight of the enemy build-up seems to be focused on trying to cut the supply lines of the armoured force and to break through into your infantry advancing along the coastal strip.

Spotter aircraft also report that the Trieste Division, a motorised force that had been held in the north, is on the move and is taking up positions along the railway track to the west of the battlefield vacated by the German 90th Light Division.

So far the 10th Armoured, holding the Kidney Ridge area and the front of Miteiriya Ridge, has seen little in the way of action.

If you want operations in the north to continue as they are for now then go to **Decision 59**. If you want to commit the 10th Armoured to the operation then go to **Decision 102**.

98

1200 hours 25 October 1942

The 1st Armoured and the 9th Australians prepare to attack as soon as it is dusk. They will be supported by elements of the 51st Infantry Division.

Throughout the afternoon the artillery readies itself to lay down a massive bombardment around the positions held by the 164th and the 7th Bersaglieri. It will be the role of the 1st Armoured to cut in behind the enemy positions and drive for the coast. They will have to also protect their flanks from possible intervention from the 15th PZ Division.

You set H hour as 2100 for the bombardment. Along the rest of the northern front you determine that there will be considerable aggressive activity to draw away any possible reserves and to confound the enemy.

With everything ready for the assault you wait until H hour. Now go to **Decision 56**.

99

0400 hours 26 October 1942

News is reaching you that the armour is coming under attack from elements of the 15th PZ Division. There are unconfirmed reports that the enemy has also committed parts of the German 90th Light Division to the attack. The Australians have made excellent progress; on the coast the Italians have fallen back and the town of Sidi Abd el Rahman is now clearly in sight, illuminated by star shells. The major conflict you were hoping to induce has finally happened. The enemy is pouring in reserves.

If you wish to commit the 10th Armoured to the fight then go to **Decision 13**. If you think the forces you have are sufficient then go to **Decision 57**.

100

0800 hours 26 October 1942

Sidi Abd el Rahman has fallen to lead elements of the Australian Division. Their positions are being protected by the 1st Armoured Division. Further south the 10th Armoured Division is fighting the 15th PZ Division and elements of the 21st PZ Division, which has been moved up from the south. This now means that the southern part of the front is only protected by the Ariete Division, as a reserve. The enemy is being forced to commit their armour to stop your progress in the northern sector.

If you think it is time to move the 7th Armoured up from the south to consolidate and exploit your gains then go to **Decision 110**. If you think it is premature and that an opportunity might present itself further south then go to **Decision 152**.

101

0800 hours 26 October 1942

The 1st Armoured Division is now protecting the Australians, who have managed to capture Sidi Abd el Rahman in the north.

The 10th Armoured Division is under considerable pressure. Not only is it fighting off attacks from the 15th PZ Division, but it is also reportedly encountering elements of the 21st PZ Division. This means that Rommel has shifted the last of his German reserves from the south and only the Ariete Division remains as an armoured backup for the troops on that part of the line.

It might be just the time to move your own armour from the south, to reinforce your guns in the north. If you want to move the 7th Armoured Division then go to **Decision 110**. If you think it should remain there for now then go to **Decision 152**.

102

0900 hours 26 October 1942

Your gamble to commit the 10th Armoured seems to have paid off. They are now in action against the 15th PZ Division and, reportedly, lead elements of the 21st PZ Division. Rommel has clearly decided to commit his last German armoured reserves to the fight in the north.

Your 1st Armoured Division has taken Sidi Abd el Rahman in the north. It is providing a protective armoured shield around the town as the Australians move in to occupy it.

You feel you need to take advantage of the gains, but are unsure about your next move. If you want to move the 7th Armoured from the south to reinforce the northern effort then go to **Decision 110**. If you think the 7th might be useful to exploit opportunities in the south then go to **Decision 152**.

103

0600 hours 26 October 1942

Horrocks, the corps commander, has committed the British 131st Infantry Brigade to support the Free French. The additional pressure and firepower has worked to your advantage. A gap in the Italian line has developed, stretching from the El Taqa Plateau through to Qaret el Himeimat. Reports suggest that the Pavia Division is in full retreat and the lead elements of the 7th Armoured

Division have reinforced the French and the 44th Infantry and have reached the edge of the El Taqa Plateau. Now turn to **Decision 104**.

104

1000 hours 26 October 1942

Harding's 7th Armoured Division is now fighting a tank battle around the eastern edge of the El Taqa Plateau. The enemy, mainly Italian tanks of the Ariete Division, is fighting back, but their tanks are no match for your armour.

It seems that the Pavia Division is all-but broken and is giving ground; clearly they fear encirclement, which is a distinct possibility. You need to keep up the momentum of the assault. It is likely that the enemy is trying to re-establish a line further to the northwest, perhaps around the Jebel Kalakh area. If you think that the 7th Armoured is sufficient as your main cutting edge in the south then go to **Decision 187**. If you want to consider shifting another armoured division from the north to support them then go to **Decision 216**.

105

1600 hours 26 October 1942

It is not long before the dust clouds warn your tank crews that the enemy armour is advancing towards their position. The cutting edge of the 21st PZ is the 5th PZ Regiment, predominantly equipped with Pz IIIs. They are moving rapidly now, barely a mile from your positions.

Your artillery opens fire; the enemy responds with their own artillery and smoke, which lands around your forward positions. Over the radio you can hear the crack and bark of tank guns firing off. The next few minutes will be crucial and you hope that the 7th Armoured is sufficiently prepared for the encounter. Now turn to **Decision 19.**

106

1500 hours 26 October 1942

You give Horrock's opinion careful consideration and come to the conclusion that you need to establish a solid reserve to exploit opportunities in the north. It will mean thinning out your line in the north and moving up the 7th Armoured Division from the south. This will mean that the southern part of your army will be fighting a holding and probing action for the time being. You pull the New Zealand Division and the 10th Armoured Division out of the line in the north and tell an excited Horrocks to move the 7th Armoured north. Now go to **Decision 153.**

107

0600 hours 27 October 1942

Your new line stretches from the El Taqa Plateau to the Munassib Depression. The new enemy line seems to be establishing itself, hinged on Jebel Kalakh. During the night elements of the 21st PZ Division tried to probe your new defensive line, but were beaten back.

Although you have achieved much, the bulk of the enemy in the north is still intact and their supply lines are relatively secure. You will need to make decisions about your course of action. If you wish to wait for developments here in the south then go to **Decision 150.** If you feel it better to attack now then go to **Decision 149.**

108

0600 hours 27 October 1942

Horrocks' war of attrition against the 21st PZ Division and the Ariete Division has ended in failure. Remnants of the 7th Armoured have been forced to retreat. With the attempt in the south a failure and operations in the north at a standstill you have no alternative but to contact Churchill and tell him of your predicament. Now go to **Decision 234.**

109

0200 hours 28 October 1942

'Swing the 10th Armoured to support the 7th and catch the 15th PZ in a pincer movement,' you shout down the telephone to Horrocks, waving your hand across the map on the wall.

'Yes, consider it done,' he replies.

'Meanwhile,' you begin, 'unleash the 1st Armoured Division against the enemy in the north. I think we've got him now,' you tell your chief of staff.

Now turn to **Decision 151**.

110

1900 hours 26 October 1942

With the fighting continuing around Sidi Abd el Rahman, and mounting as the German 90th Light Division is thrown into the combat, your 1st Armoured Division is forced onto the defensive.

Slightly further south the 10th Armoured Division, now fully supported by the New Zealand division, is holding off concerted German counterattacks. The 7th Armoured has finally arrived in the north, but will not be ready to get underway until at least midnight. The question is, where is it best to commit them? If you want to use them to support the 1st Armoured Division then go to **Decision 119.** If you feel they are needed to help the 10th Armoured Division then go to **Decision 161**.

111

0800 hours 27 October 1942

It takes a considerable period of time for your troops and armour to reorganise. The lacklustre display, particularly from your armour, so far is cause for concern. You feel that the battle needs to be fought in the north. This is the pivotal sector of the battlefield. What is needed is a fresh approach to draw the enemy armour into a killing zone, where your weight of numbers in tanks, anti-tank guns and artillery can cripple them as an effective fighting force. By destroying his mobile troops the rest of Rommel's army will be at your mercy. They will find it impossible to extricate themselves. Now go to **Decision 219**.

112

2300 hours 27 October 1942

The Rifle Brigade is tasked to take Snipe and the 2nd Battalion of the King's Royal Rifles to deal with Woodcock. If they are taken then armour will pass around the north of Woodcock and the south of Snipe at dawn.

You can see from the reports that this is going to be a hard fought contest. Both battalions have difficulty in finding their way to their targets and by dawn the Rifle Brigade is in position, but the King's Royal Rifles are well short of their objective.

If you want to commit the armour then go to **Decision 194**. If you feel the operation is a failure and wish to order them back then go to **Decision 26**.

113

2400 hours 28 October 1942

With your force in the Woodcock and Snipe areas taking up a defensive posture, in the north the Australian troops, with tank support, are ordered to push out from the Point 29 area.

The 20th Australian Infantry Brigade, supported by the 40th Royal Tank Regiment, push northwest. This should allow the 26th Australian Infantry Brigade and the 46th Royal Tank Regiment to strike northeast to take a location known as Thompson's Post. From there they will be able to advance on the rear of the enemy in the coastal area.

The 20th manages to take its objectives, but the 26th runs into trouble; the infantry are forced to ride on the Valentine tanks of 46th Royal Tank Regiment to cover the ground. They run into the 125th PZ Grenadier Regiment and the 7th Bersaglieri Regiment. The advance soon comes to a halt and is in chaos.

If you want the force to reform and continue their advance then go to **Decision 195**. If you think that insufficient hours of darkness remain to do this then go to **Decision 27**.

114

2400 hours 30 October 1942

Rommel has been throwing elements of the 21st PZ Division at the area around Thompson's Post. He has temporarily managed to make contact with the virtually cut-off 125th PZ Grenadier Regiment.

The road- and railway-crossing have changed hands several times, but for the most part it is in no-man's land. This is a deadly area that is littered with the bodies of Germans and Australians and burning tanks and vehicles.

You hear the brilliant news that two more of Rommel's supply ships have been sunk by torpedoes off Tobruk. You feel that one more major push will see Rommel forced to withdraw or suffer absolute disaster.

Patiently you wait for the enemy to dash itself on your defensive line, slowly building up your reserves and preparing for your own reserved offensive. You receive a call from Churchill. He is concerned that little progress has been made. If you agree with him about the lack of progress then go to **Decision 234**. If you feel you are on the verge of victory then go to **Decision 72**.

115

1800 hours 31 October 1942

The question is where to attack? Your objective has to be to destroy the enemy armour, make him fight in the open, and take control of his supply routes. If the attack is to be successful, then aerial sorties and clear lines of attack need to be established now. If you want to change your mind and postpone for twenty-four hours then go to **Decision 157**. If you want to press ahead then you have two choices. You can attack along the coastal strip in the north; go to **Decision 196**. Or you can strike further south, with the objective of Tel el Aqqaqir, to the northwest of the Kidney Ridge feature on the lateral supply track to Sidi Abd el Rahman. If you prefer this option then go to **Decision 222**.

116

0800 hours 1 November 1942

The attack, ill-conceived and lacking in support, has understandably failed. You were too premature and had not taken the time to use all of your resources to the full.

You determine that you will hold back and re-launch the offensive tonight. The day needs to start well with the maximum use of your resources. If you think that it is too late to guarantee any victory then you need to tell Churchill and go to **Decision 234**. If you want to press on with your planning and hit Rommel to the south of Point 29 with everything you have then go to **Decision 73**.

117

1900 hours 26 October 1942

The shattered remains of the German and Italian tanks and vehicles lie scattered across the battlefield. Huge craters dot the landscape and added to this there are clear and obvious paths through the minefields. This could be the opportunity that you had hoped for; perhaps a chance to renew your own offensive and push on to finish Rommel and his forces. Although your front line troops are exhausted, your losses have been comparatively light. Any casualties inflicted on Rommel's forces are irreplaceable for him and you feel content that you again have the initiative on your side. To consider your options go to **Decision 159**.

118

1400 hours 27 October 1942

Progress continues to be promising, but aerial reconnaissance confirms that huge columns of retreating Italian and German troops are making their way west, towards the Telegraph Road. The retreat is being covered by the Trieste Motorised Division and elements of the Ariete Division.

On the coastal strip the 15th and 21st PZ Divisions, now seriously depleted, are still holding a thin line along the Rahman Track. This has allowed large numbers of the 164th Division to escape to the west. In places large numbers of Italian troops, out of ammunition, food and water, have given up the fight and are marching placidly into captivity. This is holding up your infantry divisions, as they struggle with the logistical nightmare.

You begin to worry whether you should have been more aggressive; too many of Rommel's troops are slipping away to the west. They will only retreat for a time; you know they will turn and fight again, this time with shorter supply lines whilst yours are becoming significantly longer. Now turn to **Decision 199**.

119

0100 hours 27 October 1942

The 7th Armoured slips through the cleared lanes in the minefields and reaches Sidi Abd el Rahman. They will take over the front line duties from the 1st Armoured, in order to allow them to pull back, reorganise and refit.

Reports suggest that the Italian Trieste Motorised Division has moved east, as a backup for the 90th Light Division. You wonder whether this is likely to be a counterattack, or whether the intention is to cover a retreat.

Further south the pressure continues to mount against the 10th Armoured Division. Casualties are reported to be high, but they have inflicted irreparable losses on Rommel's remaining armour. You know that the next few days will be crucial. Now go to **Decision 200**.

120

0715 hours 2 November 1942

'Pull them back!' you order. 'Get them out of there before it's too late!

You slump down in a chair, exhausted and mentally finished. You have lost your nerve at the crucial time. The 9th Armoured Brigade should have been sacrificed in order to ensure victory, but at the last possible moment all their losses have been for nothing.

Despite pulling them out, of the ninety-four tanks with which they began the attack only ten make it back to the start line. You know that you have to break the bad news to Churchill. He will not like what you have to tell him. He wants victory not excuses. Now go to **Decision 234**.

121

1130 hours 2 November 1942

Your 1st Armoured Division rolls forward to engage the enemy 15th, 21st and Littorio formations. Overhead the desert air force gives intense support, shooting up soft skin vehicles and pinning down supporting infantry. Your artillery makes it impossible for the enemy armour to advance in a cohesive group. Your well-placed anti-tank guns protect from any flanking moves.

Finally you have the major engagement of armour that you have been seeking. This is a chance to destroy the last of Rommel's tanks. You know this is going to be a critical time. But most crucially you need to ensure that you dominate the battlefield to prevent Rommel from salvaging any damaged or broken down vehicles. Now go to **Decision 35**.

122

1130 hours 2 November 1942

The 1st Armoured Division bursts forward, through the dwindling anti-tank screen and engages the enemy armour.

You feel you need to direct operations yourself and quickly head for the salient, determined to see your tanks in action and to ensure they exploit every possible advantage.

The 1st Armoured are fully engaged by the time you arrive and the gun flashes are indistinct in the swirling clouds of sand.

'Reports!' you bark.

Staff officers and commanders scurry to find out the latest news. You stare impassively through your field glasses, expertly noting each move of your regiments and the slackening of fire.

'They're breaking and heading west,' shouts a staff officer. 'There's at least fifty or more tanks knocked out or abandoned on the battlefield!'

'Excellent!' you reply.

Now it is time to enlarge the gap in the enemy line and ensure that the 7th Armoured has every opportunity to break through and finish this. Now go to **Decision 163**.

123

0800 hours 3 November 1942

The 2nd Armoured Brigade has struck at the north of Tel el Aqqaqir, the 8th to the south of the town, and the 7th Armoured Division directly at the town.

By daylight it is clear that a huge gap has been opened up in the enemy defensive line. It appears that the remnants of the 21st PZ Division are retreating with some haste to the northwest. The heaviest fighting is to the south of the town, where the Ariete Division has dug in with their remaining armour and simply refuses to budge. You now have to determine whether to pursue or to consolidate. If you want the 2nd Armoured Brigade to give chase to the 21st PZ Division and for the 7th Armoured Division to forge on northwest then go to **Decision 82**. If you feel that the priority needs to be to destroy the Ariete Division then go to **Decision 125**.

124

2200 hours 3 November 1942

The 151st Infantry Brigade begins their advance on the new enemy defensive line. It is pounded by your artillery, although it is difficult to judge whether the bombardment is effective.

Suddenly the Germans launch an unexpected counterattack. The signals from the front are confused, but it is clear that it is German armour, probably the 21st PZ Division. Rommel has ordered a surprise attack and your infantry lacks the protection of your own tanks. They will have to deploy quickly or retreat if they are to avoid being overrun. If you want to order them to pull back then go to **Decision 166**. If you want them to hold and send the 2nd Armoured Brigade to help them then go to **Decision 204**.

125

0800 hours 4 November 1942

The major battle to dislodge and destroy the Ariete Division has taken virtually twenty-four hours to accomplish. By the end of the action the battlefield is littered with burned out Italian tanks and the area is covered with the dead. After putting up a strong resistance to the overwhelming attack the Italians have begun to surrender in huge numbers. Often they have fought to exhaustion, lacking ammunition to continue the struggle.

The fight has also led to the elimination of the Littorio and Trieste Divisions; they have virtually ceased to exist as fighting units.

Your reconnaissance aircraft report that the remaining Italian infantry divisions, notably the Bologna and Trento, have been streaming west in a desperate attempt to escape. Lacking transport, food and water they too have been surrendering in increasing numbers. Now go to **Decision 39**.

126

0800 hours 5 November 1942

Although the progress is disappointing you now impose even greater objectives for your forward units.

The 7th Armoured will go cross-country to cut the coastal road at Sidi Haneish, some sixty-five miles to the west of the Rahman Track. The 1st Armoured, now to the west of El Dada, is to take a broad detour to Bir Khalda, some eighty miles west of the Rahman Track, and then attempt to cut the road at Mersa Matruh.

You now hear that the fuel supply for your forward units is becoming a major problem. If you want the armoured divisions to press on and hope that they can be refuelled over the course of the day then go to **Decision 205**. If you would prefer to ensure that the 1st Armoured gets priority supplies then go to **Decision 228**.

127

0800 hours 5 November 1942

With the New Zealanders en route along the coastal road to Mersa Matruh, the problem with your armour is that it is running out of fuel. The 1st Armoured is heading for Mersa Matruh, but reports are that fuel is extremely low and that the fuel trucks are some distance behind them. The 7th Armoured is still moving towards Sidi Haneish, via Bir Khalda. The

10th Armoured Division is largely on the coastal road and has ample fuel supplies.

You determine to try to refuel the 1st and 7th Armoured as quickly as possible and to push the 10th Armoured as far down the coastal road as possible. Now go to **Decision 208**.

128

0800 hours 6 November 1942

The supply trucks have finally reached the 1st Armoured Division, but only with sufficient fuel to resupply two armoured regiments. These set off towards Mersa Matruh.

By midday a heavy rainfall has begun, turning the dirt roads into quagmires. This makes the problem of resupplying your lead units all the more difficult. The supply trucks are still some forty miles short of the 1st Armoured Division.

The New Zealanders are slowly making their way towards Sidi Haneish and the 10th Armoured has sprinted towards Fuka. Meanwhile, the 7th Armoured, now to the southwest of Sidi Haneish, has encountered elements of the 21st PZ Division and a German reconnaissance group. If you want to order the 7th to engage then go to **Decision 230**. If you want them to flank the Germans and press on to the west then go to **Decision 43**.

129

0700 hours 5 November 1942

With barely twenty operational tanks, the remnants of the 15th PZ Division, supported by infantry and some artillery, have been in action against Freyburg all night.

Leaving six tanks burning in the desert, the bulk of the force has slowly slipped around Freyburg and made off to the west. The only hope is that the night manoeuvres have depleted the enemy's fuel and that they are not resupplied before they can reach safety. You are still well ahead of schedule, but during the night the primary focus has been on ensuring that the armoured divisions are refuelled. Now go to **Decision 169**.

130

1200 hours 6 November 1942

You decide to call off the pursuit, content that you have inflicted serious damage to Rommel's army. The only thing that irks you is the fact that he has been able to extricate the bulk of his German troops. You know he is down to the bare bones in terms of equipment, but Rommel is resourceful and will, no doubt, fight again when the time is right.

You have managed to kill, wound or capture around 37,000 of the enemy, the vast bulk of them being Italians. As for the German core of the army, they have been reduced to around 5,000 men and twenty tanks.

You were caught by surprise with Rommel's rapid retreat, combined with the fact that the corps were slow in their pursuit. Your failure to cut Rommel off at Fuka and Mersa Matruh may well prove to be costly in the long run. As Churchill would later say, perhaps the victory at El Alamein is the end of the beginning, not the beginning of the end, of the war.

131

1800 hours 6 November 1942

With heavy rain the resupply problems for your forward units are making life incredibly difficult. The Germans have also dug in around Sidi Barrani, some eighty miles to the west of Mersa Matruh. It is clear to you that Rommel intends to hold the position for long enough to allow as many of his men and their precious equipment to get through the passes at Sollum and Halfya.

Unfortunately there is little you can do about it. Your armoured units are at the very end of their resupply chains. The last thing you want to do is to risk unnecessary casualties by overstretching yourself.

Your victory on the battlefield is complete, yet you feel cheated because Rommel has managed to extricate vital elements of his army. Pursuit will resume, but you will need several days' reorganisation before that can happen. In the meantime Rommel will continue to fall back to the west. You know you will face him again in battle.

132

Miteiriya Ridge seems to be a scene of chaos. With the exception of two points all of the objectives given to the infantry have been taken. In the 51st Infantry Division's sector, the end of the northern corridor is being covered by enemy anti-tank guns. This is the exit route for the 1st Armoured Division. By now they should have established themselves around Kidney Ridge, to the west.

In the far south elements of the 1st South African Division are still 500m from their objectives. In the centre, although the New Zealanders have taken up their positions along Miteiriya Ridge, the 10th Armoured Division seems reluctant to push forward to take up their covering positions further west.

If you want to focus on the northern corridor then go to **Decision 90**. If you want to focus on getting the 10th Armoured Division through in the south then go to **Decision 173**.

133

1600 hours October 24 1942

You receive a signal that informs you that the 10th Armoured Division, supported by the corps artillery, are attacking across Miteiriya Ridge. The bad news is that in the north very little seems to have been achieved. Reports from the 9th Australian Division and the 51st Infantry Division seem to indicate that the 1st Armoured Division could have broken through by now with a little drive and determination.

The whole situation is maddening; every moment wasted in indecision and lack of will to drive forward gives the enemy an opportunity to recover from the shock of the operation. You can imagine the enemy even now contemplating the moving of their own armoured forces to blunt any attempt by your tanks to break out of the advanced objectives that you have taken. All of the planning and the sacrifices will be lost. If you feel your commanders know what is expected of them, turn to **Decision 91**. If you feel you need to impress upon Lumsden and his X Corps the importance of offensive eagerness, then go to **Decision 134**.

134

'Lumsden, no word from you. What is happening?' you demand.

'The situation is unclear at the moment, but I am expecting reports to come in soon,' he tells you.

'I need not impress upon you the urgency of the situation I hope,' you tell him.

'I understand,' he replies.

'Should your commanders fail to show the drive and determination necessary then I will replace them. You must drive them Lumsden,' you roar.

'I understand. We're mounting operations in the northern and southern corridors now,' he states.

Now turn to **Decision 233**.

135

You are still staring at the situation map when the phone rings beside you.

'Montgomery," you answer.

It is the Prime Minister. He wants a situation report. You advise that the operation is around twenty-four hours behind schedule. He listens in silence as you quickly update him on the progress so far.

'So it is the armour that has let you down then?' he concludes.

'To some extent Prime Minister,' you reply.

'So what are you going to do about it? I want a victory. The Empire needs a victory Montgomery, do I make that plain?'

'You will have your victory I hope,' you tell him, although he does not seem that convinced.

If you want to try and reassure him then go to **Decision 213**. If you feel you have said enough then go to **Decision 234**.

136

'I need to make myself clear,' you begin. 'Your corps will get out west of the minefield where it can manoeuvre. You are to locate and destroy the enemy armoured battle group. You are to make sure that the operations of the 2nd New Zealand Division, southwest of the Miteiriya Ridge, are not interfered with by enemy tank forces.'

You need this to happen in order to ensure that the New Zealanders can begin their crumbling work south-westwards, using the 9th Armoured Brigade as support. Meanwhile, the armour will form a hinge and a block to prevent enemy tanks from getting involved in the fighting to protect their own infantry. Now turn to **Decision 94**.

137

With the 1st Armoured in a more forward position you consider shifting the axis of attack. Perhaps a new opportunity has presented itself. The 1st Armoured can protect the flank of the Australian 9th Division. They can begin crumbling operations against the enemy in the north, between you and the coast. The presence of the 1st Armoured can achieve precisely the same objectives if it threatens the main lateral supply route of the enemy army, along the Rahman Track. The rest of your northern front can dig in and release reserves for this effort. You call a conference at midday.

The Australian commander, Morehead, tells you that he will be ready to make an attempt tonight. You decide that this is the best course of action and that the enemy will have to respond to this threat. It will take close cooperation between your Australian units and the 1st Armoured but hopefully this can be achieved.

Now turn to **Decision 98**.

138

The options open to you are clear. You can either hope that the Germans commit their armour in the north and that you defeat them, which will force the northern movement of their last armoured reserves, or you can attempt to force the decision on them. It seems that the following are possible: if you want to send your armour west then go to **Decision 180**; if you want to swing the 1st Armoured to the north and coordinate an attack with your two infantry divisions on the coast road then go to **Decision 9**; if you want to shift your Greek Brigade south to support the defence (leaving the 50th Infantry Division on a broader front) then go to **Decision 53**.

139

To the north the Australian Division, as part of the crumbling operation, are to move into position to attack Point 29, an Axis artillery observation post to the west of Tel el Eisa. The capture of this position will aid in the surrounding of the Axis salient on the coast. It will directly threaten the German 164th Division and Italian units.

In the afternoon the Axis forces probe your defensive armoured line for weaknesses. Your men are well prepared and the enemy find nothing they can exploit.

As the sun sets it is time for your infantry to begin operations again. If you want to support the Australian's attack on Point 29 with artillery and armour then go to **Decision 11**. If you feel that they can take it on their own then go to **Decision 214**.

140

1900 hours 25 October 1942

It takes the bulk of the day for the units to shuffle south to cover the gap in the line vacated by the Greek Brigade. It is a worry that your centre is so thinly held, but it is also thinly held by Italian units on the enemy front. You need to be able to be in a position where the armoured units of the 7th Armoured Division can punch through.

The Greek Brigade moves up to face the 185th Folgore Division, leaving you with the entire 44th Infantry Division and the Free French to achieve this. Your men wait nervously until it is dark. With the Qattara Depression to their south at least their flank is secure. It is now a question of how far you wish the units to advance if the attack is successful.

If you wish to set the objective as the Qaret el Himeimat, on the front line of the Pavai Division's defensive position, go to **Decision 144**. If you wish to push through beyond that and take the El Taqa Plateau, beyond the Italians, then go to **Decision 185**.

141

The enemy's 164th Division is still firmly entrenched and is causing huge casualties amongst the advancing Australians. Point 29 still serves as a major obstacle to the effort. Artillery fire nearly obliterates the view of the German positions, yet unnervingly accurate return fire continues to suppress the lead elements of the Australians.

The 1st Armoured has nearly broken through the crumbling front of the 164th Division. There is disquieting news, however. Elements of the 15th PZ Division, supported by lead units of the German 90th Light Division, are reported to be engaging the flank of the armour. At last the enemy has committed his reserves.

If you want to press on with the attack then go to **Decision 100**. If you feel you should pull back your armour to safety then go to **Decision 142**.

142

0800 hours 26 October 1942

By sunrise the bulk of your forces are back in the same positions they began in. The armoured units have suffered minor casualties, but the crews are exhausted. The offensive has been a comparative failure and the ground you have gained is minimal.

If you want to contact Churchill and tell him you fear that victory is eluding you then go to **Decision 234**. If you want to reconsider your options then go to **Decision 24**.

143

1500 hours 26 October 1942

The 15th PZ Division and elements of the 164th Light Division have been attacking your positions in the north. Your heavy artillery has broken up the offensive and casualties have been minimal.

Reports seem to suggest that Rommel is beginning to move his armoured reserves up from the south. The 21st PZ Division and part of the Ariete is on the move. You have ordered your air force to punish them for moving during the day.

You have made some attempts to consolidate around the Kidney Ridge area, but German anti-tank guns have largely negated your efforts. You are deeply concerned that your offensive is on the brink of breaking down. Now go to **Decision 111**.

144

0200 hours 26 October 1942

Initial reports are promising; the Free French have penetrated the defensive minefield on the far southern edge of the Pavia Division's defensive line. They are engaging the enemy infantry at close quarters and gaining ground. The main effort by the 44th against the bulk of the Pavia's line has stalled, largely due to a thicker-than-expected minefield. Casualties are light, but the progress is painfully slow. With only a few hours of darkness remaining you had hoped for better overall progress than this. The Italians are determined and doggedly holding on. If you think the Free French can achieve the objective then go to **Decision 60**. If you want to reinforce them with elements of the 44th Division then go to **Decision 103**.

145

0600 hours 26 October 1942

Harding's 7th Armoured Division has been engaged for the past two hours. The bulk of the force has passed through your newly-gained ground and swung behind the Italian defensive line. This has caused huge panic for the enemy; in desperation they have thrown elements of the Ariete Division at your encircling tanks.

The Italian tanks are no match for your armour and have been severely mauled. News suggests that the 21st PZ Division is moving to attack and they should be in a position to mount a major offensive within the next hour or so. The opportunity to roll up the enemy line in the south all hinges on events in the next few hours. Now go to **Decision 146**.

146

1000 hours 26 October 1942

Progress has been good. The Ariete Division has pulled back and the Pavia Division is abandoning their defensive line. This has left the Folgore Division's flank exposed. The 21st PZ Division has definitely been spotted moving around the Jebel Kalakh feature to the northwest. If you want to order the 7th Armoured to move up and engage them then go to **Decision 62**. If you feel you are better served by setting up a defensive armoured line and waiting for the attack then go to **Decision 105**.

147

1400 hours 26 October 1942

You know it will take some time to shift an entire division along the Springbok Track that lies to the east of your front line. It seems that the 10th Armoured Division, commanded by Gatehouse, is the best option. Temporarily, it will be under the command of Horrocks and attached to XIII Corps. You hope that the division will be ready to take up offensive actions tonight in the south. Now go to **Decision 63**.

148

0600 hours 27 October 1942

You decide to do both. You send the 7th Armoured Division forward to break up the Italian resistance and you also decide to commit the 10th Armoured Division to the south.

It arrives just in time to work its way through the gaps in the minefield to the far south and slot in alongside the 7th Armoured.

The sudden moves seem to have taken the enemy by surprise. The Italians begin to withdraw to the northwest. By daybreak you have control of the El Taqa Plateau and your units report back that the whole of the Pavia Division is in retreat and that the Folgore Division has also pulled back.

From the plateau it is possible to see the enemy retreating across the desert, towards the Jebel Kalakh area. If you want to hold your troops and consolidate these gains then go to **Decision 190.** If you want to immediately order them to advance and take advantage of the situation then go to **Decision 218**.

149

1000 hours 27 October 1942

For the past few hours a major tank battle has been raging around the Jebel Kalakh area. The enemy has committed both the 21st PZ and the remnants of the Ariete Division. Horrocks reports that the 7th Armoured has broken through on the left flank and has reached the Axis supply road, leading northeast. The 10th Armoured, working alongside the 44th Infantry Division, has pushed north and has run into elements of the Breschia Division. It seems that the bulk of the enemy forces are now falling back parallel to the Ruweisat Ridge. Scattered units have set up blocking positions along the supply routes, but resistance is relatively light.

The opportunity to finish the enemy is close. But you will need at least twenty-four hours to realign your units and prepare for another major assault. You fear that your forward units are too exposed. Now turn to **Decision 191.**

150

1400 hours 27 October 1942

It is now a question of whether you will resume the offensive or wait for the enemy to make a move. Clearly he has been defeated to some extent in the south, with several of his major units mauled and disorganised. The fact remains that the bulk of the army is still intact in the north. Operations there have continued at a lower intensity, but perhaps it is now time to refocus on the north? Now turn to **Decision 219**.

151

0800 hours 28 October 1942

Flying over the battlefield you can clearly see your armoured units advancing on all fronts. The 7th and 10th Armoured have crushed the 15th PZ Division and wreckage lays strewn across the desert. You can see lines of enemy infantry trudging east into captivity. Over the northern sector the main enemy defensive lines have been overrun. At 0530 the 1st Armoured Division punched its way into Sidi el Rahman. This caused a huge panic in the enemy lines. Despite a counterattack by the German 90th Light Division the armour held until the 9th Australian Infantry joined them and set up a hedgehog defence around the town.

The enemy armour, deprived of its fuel and supply, has been trying to escape to the west all through the night. Thousands of enemy infantry have been cut off. Your commanders report low casualties, but difficulties in dealing with so many prisoners.

'Gentlemen, this is a complete and absolute victory!' you exclaim. 'Battles never quite go as planned, but victory it is.'

Over the next two days you ruthlessly pursue the enemy, but Rommel is wily and manages to extricate as much as he can. You will face him again one day very soon. The road to Tobruk is open, but you know Rommel will turn and fight when he thinks he is ready.

152

1200 hours 26 October 1942

After careful thought you order Horrocks to launch a major offensive in the south. You suspect that elements of the 21st PZ Division are still in the south, but certainly a reasonable part of the formation is now operating in the north.

Horrocks is delighted to be able to get going; his men are desperate to get into action and always feared that their presence in the south would mean that they would only ever be in a sideshow in the battle. He tells you that the offensive can get underway in a matter of hours. His plans are clear and his men ready for action. Now go to **Decision 62**.

153

1800 hours 27 October 1942

After enormous deliberation you determine to focus on the north and make the best use of your new reserve force. The problems are enormous; the logistical headaches of shifting men and equipment in the desert are taxing your commanders to the point of exhaustion. All of this is going to take a huge amount of time and you can only hope that Churchill is patient with you, as you prepare for the next phase of the offensive. Each time you plan for the renewal of the offensive another problem prevents you from carrying it out. The days pass with sporadic fighting along the whole front, particularly in the north. No gains of any significance are being made and the enemy is still as resilient as ever. Now turn to **Decision 28**.

154

2300 hours 27 October 1942

You opt for Snipe as your solitary target, allocating the Rifle Brigade to take it and then at dawn the 24th Armoured Brigade will move in to support them. The battalion finds it hard to reach its positions by dawn, but the shell bursts from your supporting artillery does help them find their way. The 24th Armoured Brigade sets off to cover the Rifle Brigade, but immediately runs into tanks belonging to the Italian Littorio Division and elements of the 15th PZ Division. Your armour is kept away from Snipe and it seems that Rommel is about to launch a major counterattack against the position. Now go to **Decision 70**.

155

2400 hours 28 October 1942

The 20th Australian Infantry Brigade and the 40th Royal Tank Regiment push forward northeast from the area around Point 29. From there the 26th Australian Infantry Brigade, with the 46th Royal Tank Regiment, is to take over and push northeast towards Thompson's Post. They will then cross the railway to the coast road and attempt to cut the enemy off. An attack can then be launched against the enemy pinned against the coast from the southeast.

The 20th achieve their objectives, but the 26th is held up by a determined counterattack. With darkness beginning to be lost, you wonder if they can reorganise in time and push forward again.

If you want to call off the second phase of the operation then go to **Decision 27**. If you want to order them to reorganise and press on then go to **Decision 195**.

156

0800 hours 31 October 1942

Your hastily-organised counterattack comes under immediate pressure from elements of the 90th Light Division and the 21st PZ Division. Valentines of the 46th Royal Tank Regiment rumble forward into a firestorm of German anti-tank guns and hidden German tanks.

The lead commanders of the regiment report that the German vehicles are reluctant to move out of their positions, which may suggest they are very low on fuel. Despite this, they hardly need to do so as they are inflicting terrible casualties on your tanks.

As you survey the map and read the reports you receive a call from Churchill. He is clearly angry about your lack of progress and when you tell him about your latest counterattack he is even less impressed.

If you want to tell him that he should wait a day or two more, as you are preparing for a major offensive, then go to **Decision 72**. If you agree with him and you are concerned about the whole situation then go to **Decision 234**.

157

1800 hours 31 October 1942

You have had a tense conversation with Churchill, who questioned your decision to postpone the offensive for twenty-four hours. You managed to assure him that thorough preparations were the key to achieving success and that nothing should be left to chance.

Decisions now have to be made with regard to the point of the attack. It is clear that Rommel expects you to renew your assault in the immediate coastal sector of the battlefield. The key options are to either attack north of Point 29, where Rommel anticipates the attack - go to **Decision 29** - or to attack to the south of Point 29, with the first objective being Tel el Aqqaqir - go to **Decision 73**.

158

0800 1 November 1942

The attack has been ruinous. You had expected the 9th Armoured Brigade to take heavy casualties in attempting the breakthrough, but believed that this would be more than compensated by the fact that the 1st Armoured Division would cause chaos in the enemy rear areas and smash any armour that Rommel had been able to muster. Unfortunately, by the time the 9th Armoured Brigade was in a position to attempt a breakthrough the enemy had been able to pick off all of the tanks that had dared to approach. At least twenty-four well-placed 88mm flak guns, used in an anti-tank role, were especially effective against your armour. The whole attack has ground to a bloody stop. The 9th Armoured Brigade has virtually ceased to exist and the New Zealand Division, despite reinforcement, has been severely mauled. You have no choice but to break the bad news to Churchill and go to **Decision 234**.

159

2100 hours 26 October 1942

You call a conference of all your key commanders to look at your options and consider the best way forward. The options are clear to you; clearly your commanders have their own agendas and want to ensure that their voices are heard.

The two main options are to consider a northern offensive, or to shift operations to the south. Both have their advantages and disadvantages.

In the north the bulk of the enemy is still in place, facing your own forces, but as the line continues south the enemy concentrations become thinner.

You still have considerable forces to the south. Moving troops south will mean disruption and disorganisation.

Horrocks, in the south, has not been idle. His 44th Infantry has taken up advance positions. His armour and most of his infantry are threatening the El Taqa Plateau and reports suggest that the Italians in the south are falling back to new positions hinged on Jebel Kalakh.

If you want to shift the 10th Armoured Division to the south to reinforce Horrocks then go to **Decision 63**. If you would prefer to continue to focus on the north after reorganisation then go to **Decision 219**.

160

1400 hours 27 October 1942

Horrocks in particular has made enormous progress, virtually unimpeded by the enemy and only slowed by vast numbers of enemy troops trying to surrender. His infantry units are lagging far behind, but advanced units are trying to deal with the sheer numbers of enemy prisoners.

In the north it is now clear that Rommel has given the orders for a full retreat. Rommel is only holding together his army by force of character. An intercept of a telegram from Hitler has ordered Rommel to stand and die.

Your forward units in the north report that the enemy line is being held by the 90th Light Division, supported by around twenty tanks. Your tanks are massing to the east and south, some 650 of them.

If you want to allow Rommel to stand and hope he will retreat then go to **Decision 32**. If you want to put in one last attack on his flanks then go to **Decision 76**.

161

0100 hours 27 October 1942

The 7th Armoured Division moves west to support the 10th Armoured. In a matter of hours it becomes clear that the additional tanks have made all the difference in this sector and that the German and Italian tanks have fallen back into defensive positions.

The 7th and 10th are now within striking distance of the Rahman Track. Both you and Rommel know that the possession of this vital arterial route is the key to the battle. As long as Rommel holds it he can resupply his troops in the south, or at least extricate them if necessary. If you can take and block the track then most of his troops to the south will be cut off. Most of them lack motorised transport and their escape will be impossible.

If you want to order the 7th and 10th to take the Rahman Track and press west as far as the Telegraph track then go to **Decision 224**. If you want to order them to hold and consolidate then go to **Decision 33**.

162

0715 hours 2 November 1942

Although confused, the reports from the forward observers and commanders seem to suggest that the 9th Armoured Brigade has broken through as planned. As the minutes pass you hear that over thirty-five enemy anti-tank guns have either been destroyed or overrun. The armoured brigade is shattered as a fighting force, with only twenty-four of the ninety-four tanks that started the attack still being capable of fighting. No doubt your engineers will be able to salvage some of these.

Worrying messages are beginning to reach you. Although a gap has been created and several hundred prisoners taken, the opinion is that the gap is insufficient for the 1st Armoured Division to attempt a breakthrough. This is disastrous news.

If you want to push the 1st Armoured Division forward and force the gap then go to **Decision 34**. If you feel that the 9th Armoured Brigade and artillery should be used to do this first then go to **Decision 78**.

163

1600 hours 2 November 1942

You launch your 133rd Lorried Brigade and the 151st Infantry Brigade at the Snipe and Skinflint objective, in the northern sector of your salient. The assault is preceded by yet another devastating artillery barrage.

The brigades run into unexpected opposition in the shape of the Italian Trieste Motorised Division. As they close in on the objectives, they are met with the sight of shattered transports and isolated pockets of defence.

The operation is over very quickly; the artillery bombardment has broken the back of the enemy division and casualties for your infantry are incredibly light. You now need to prepare for your next thrust. Time is on your side to make all of the organisational changes you would want. If you wish to focus on getting the 7th Armoured ready for action during the night then go to **Decision 226**. If you feel it is more important to assemble infantry brigades to crumble the remainder of the enemy line then go to **Decision 36**.

164

1745 hours 3 November 1942

The 151st Infantry Brigade has begun their advance, supported by the 8th Royal Tank Regiment. They are attacking enemy positions south of Tel el Aqqaqir. Throughout the day your desert air force has flown over 1200 sorties, shooting at any enemy troop concentrations, tanks and transports.

The 5th Indian Brigade is due to start in the early hours of 4 November by attacking the area around the Rahman Track, somewhat further south. At 0616 the 154th Infantry Brigade is due to attack Tel el Aqqaqir itself.

The timings of these attacks are crucial as they hold the key to the enemy defensive line. If a solid breach can be made then your armour can do the rest.

The 151st reports that is has reached its objectives and is digging in. Confused by this signal you check the coordinates and they have in fact stopped short of their objective.

If you want to order them forward to take their actual objectives then go to **Decision 202**. If you think their move is sufficient then go to **Decision 37**.

165

0730 hours 4 November 1942

It is time for the pursuit. You decide to launch the 1st and 7th Armoured Divisions to swing northwards and roll up the enemy line.

Meanwhile, two New Zealand Lorried Infantry Brigades, along with the 9th Armoured and the 4th Light Armoured Brigades, will head west to take the escarpment above Fuka, some sixty miles to the west.

It is crucial that your troops keep moving to trap as many of Rommel's units as possible and prevent them from escaping. You expect there to be pockets of resistance here and there, but your hope is that Rommel's forces will be more intent on making for safety than turning to fight.

The first news you receive comes several hours later. The New Zealanders took longer to get started on their pursuit due to the fact that the units were dispersed. They then ran into minefields and the narrow tracks between them have become congested. Freyberg, their commander, reports that he is just fifteen miles west of the Rahman Track and it is getting dark.

If you want him to stop and resume in the morning then go to **Decision 227**. If you want him to continue then go to **Decision 237**.

166

0600 hours 4 November 1942

A bloody night-time battle has seen the 151st Infantry Brigade forced to give ground in the north. They have suffered unnecessary casualties and for a time the German armoured thrust threatened to expose the flank of your salient. Nonetheless you were shocked to hear that the German armour seemed unwilling to stray very far from the Rahman Track.

There has to be a reason for this. Perhaps this is part of Rommel's strategy of containment and counterattack, or maybe it points to the fact that his fuel supply situation is more critical than you had thought.

Elsewhere, the 5th Indian Brigade has taken its objectives and the Argyll and Sutherland Highlanders, at the head of the 154th Brigade, have just marched into Tel el Aqqaqir. Latest news suggests that the 21st PZ Division, having forced the 151st back, is now retiring from the Rahman Track. Now go to **Decision 165**.

167

2200 hours 4 November 1942

The minefield turns out to be a dummy, hastily laid to try to fool you. The 15th PZ Division attempts to stop your advance, but they are too weak. Your commanders report that the enemy seems very reluctant to manoeuvre, preferring to fall back as fast as they can. You suspect that not only are they weak in tanks but also that the fuel situation has now reached a critical point. The enemy simply cannot afford to waste a single drop.

By the end of the day the 1st Armoured Division has reached El Daba and the 7th Armoured is now at Galal. Increasing numbers of German prisoners are being reported by your units; it seems that you might be winning the pursuit.

One problem is becoming clear; you are outrunning your own supply lines. The leading armoured divisions report that fuel supplies are very low. If you want to order your divisions to halt and wait for fuel then go to **Decision 238**. If you think they should press on then go to **Decision 40**.

168

1200 hours 5 November 1942

The results of the reconnaissance flight are fairly conclusive. The crew have spotted some dummy positions around two miles from the New Zealanders and there are some scattered groups of vehicles moving slowly west, around five miles away. The observer on the aircraft, a veteran of two years of fighting in North Africa, is sure that the minefield was laid by the British some time ago. What's more, he thinks that it is a dummy minefield. If you think he might be right then go to **Decision 41**. If you think he is probably wrong then go to **Decision 85**.

169

0800 hours 5 November 1942

The fuel trucks have linked up with the 1st Armoured Division at El Daba; it will take some time to refuel. The 7th Armoured is close to Sidi Haneish, but has run into elements of the 21st PZ Division and a German reconnaissance group. It is prudent to ensure that the 7th Armoured is refuelled before they attempt to attack the Germans. Reports suggest that the Germans are slowly pulling back to the southwest of Sidi Haneish. Rain has also started falling, causing huge problems as the dust tracks begin to break up into a series of muddy rivers. The 10th Armoured has made it as far as Fuka and is fully fuelled. Now go to **Decision 128**.

170

0800 hours 5 November 1942

The 10th Armoured Division arrived at Fuka in the early hours of the morning, falling on the rear of the 15th PZ Division. The fight was fierce but brief; only a handful of enemy vehicles managed to slip away to the west. Considerable numbers of prisoners were taken, with the majority of them being German. Your decision to hold Fuka was a good one; at least another part of the German rearguard has fallen into your hands.

Elsewhere the fuel trucks have caught up with the 1st Armoured Division. The 7th Armoured is close to Sidi Haneish, where it has encountered elements of a German reconnaissance group and the 21st PZ Division. It has also begun to rain, which worries you more than the enemy. Now go to **Decision 128**.

171

1200 hours 6 November 1941

The 10th Armoured takes the lead in the race for Mersa Matruh. The 7th Armoured takes a more southerly route, hoping to work their way around the flank of the enemy should they choose to put up a fight there.

Aerial reconnaissance suggests that the Germans have already abandoned Mersa Matruh and are trying to set up a new defensive position around Sidi Barrani. The town is another sixty miles to the west of Mersa Matruh.

If you feel that you have pushed Rommel as far as is practicable for now then go to **Decision 42**. If you want to chase him towards Sidi Barrani then go to **Decision 209**.

172

1400 hours 7 November 1942

The 10th Armoured crashes through the ill-prepared defensive line at Sidi Barrani. They report that the bulk of the German troops, remnants of a mixture of different divisions, have been forced to surrender.

It is doubtful that Rommel has been able to extricate more than a fraction of his forces, added to which the bulk of his tanks and vehicles are lying abandoned on the road from Mersa Matruh to Sidi Barrani.

The 10th Armoured dare not advance any further; their fuel is low and they are overwhelmed by the numbers of prisoners that they have captured. Now go to **Decision 240**.

173

The 24th Armoured Brigade is back-up behind the 8th Armoured Brigade; a breakthrough is possible if the tanks show some dash and courage. You order the units forward. They can act as a screen for the infantry; give them time to bring up anti-tank guns and allows the focus to switch to breaking through the northern corridor. Reports suggest, however, that anything that shows itself over the ridge gets the immediate attention of enemy guns. You worry that if you do not act soon all elements of surprise will be lost. If you want to order the advance of the tanks then go to **Decision 231**. If you want to call off the advance and wait until nightfall then go to **Decision 2**.

174

You contact the commander of the 51st Infantry Division. In no uncertain terms you tell him that the blocking positions in the north need to be cleared.

'There were opportunities for the 1st Armoured to have broken through. I'm not so sure now. The enemy defences have been stiffened over the last couple of hours,' he tells you.

'You need to clear the way, is that clear? When can you get underway?' you demand.

'I'm organising things now,' he tells you. 'With a little more artillery support and some aircraft we should be able to punch through,' he confirms.

'Good. I want the 1st Armoured Division out in front of your positions as soon as possible. You are to take all necessary measures to make this happen,' you tell him.

It still concerns you that the armour seems so reluctant to get forward. Armoured divisions should be in their blocking positions now and driving in the enemy armour. You now must wait for news from the front and turn to **Decision 47**.

175

'What are our options to support the XIII Corps in the south?' you wonder, staring at the situation map.

All of the divisions running from the 4th Indian Division down to the 44th Division, covering XIII Corp's flank, are engaged. You contemplate moving one of X Corps' divisions to the south, but fear this might put your northern operations at risk. On the other hand, the armour seems to be achieving very little. You can expect a violent reaction to the guns to have been made in the north, so you will probably need all your artillery and aircraft support. Your options are limited for the time being. What worries you now is that without X Corps breaking out in the north the Germans will be able to concentrate their armour against 7th Armoured Division in the south. The more enemy armour drawn south, the weaker he will be in the north. It may mean sacrificing some of 7th Armoured for the sake of the operation as a whole. Now go to **Decision 212**.

176

0200 hours 25 October 1942

You awake to see a pale and anxious face peering down at you. It is your chief of staff, Brigadier de Guingand.

'I've been listening to the radio traffic since 2200,' he begins.

'Yes, yes and?' you urge.

'The 10th Armoured Division has not yet crossed the Miteiriya Ridge,' he continues.

'Why? What time is it?'

'It's 0200. Lumsden tells me that he is not confident that his units are set up to break out. I told him to come and see you. He hasn't, so I felt you needed to know.'

'Why hasn't he pushed forward yet?'

'It seems he's concerned that without infantry support his tanks will get picked off by the German anti-tank guns,' Guingand replies.

If you want to head for Lumsden's HQ immediately then go to **Decision 5**. If you want to call a corps commander conference in your map lorry as soon as possible then go to **Decision 49**.

177

'Lumsden I want no wavering now! Do you understand? Tell Gatehouse to get forward immediately. All depends on this,' you order.

'Gatehouse believes that pushing forward out of the ridge is suicide. He will take unnecessary casualties,' he tells you.

'I don't think you understand the situation. If deeper penetration is not made by our armour then the crumbling exercise cannot begin. Neither will the enemy be obliged to commit his armour,' you explain.

'I appreciate that, but I agree with Gatehouse,' Lumsden tells you.

If you want to concede and agree with Gatehouse and Lumsden then go to **Decision 50**. If you want to press your point then go to **Decision 136**.

178

0715 hours 25 October 1942

You understand that Gatehouse now has three regiments out, and although they are twenty-four hours late they are in a position to shield Leese's XXX Corps. The latter could now begin their crumbling operations to the left and the right.

Your confidence is high, but then you hear some shattering news. The 1st Armoured Division is not in fact on its objective line at all. The 2nd Armoured Brigade is only holding the eastern end of Kidney Ridge; the rest is occupied by German 88mm anti-tank guns. As for Gatehouse, the report of three regiments forward of Miteiriya Ridge is not true. There is no armour out in front of the ridge. There is no armour to protect the New Zealanders' crumbling operations. The news shakes you. What is more, the 7th Armoured Division in the south is not in the open ground either. It has given up its assault for now. It seems as if your whole operation is now in jeopardy. If you want to try and rectify the situation then go to **Decision 137**. If you feel that the situation is hopeless go to **Decision 179**.

179

After weeks of planning it seems that you have been severely let down by your commanders on the ground. You hope that you can come up with an alternative plan. Granted, your infantry have made impressive gains and have penetrated the bulk of the enemy minefields, but your armour has been woefully inadequate for the job. Perhaps you asked too much of them? Perhaps you were over optimistic? You reject both of these questions. It was the lack of dash from your armoured commanders and you blame Lumsden for the bulk of the failures. You contemplate your own position and wonder what to do. If you wish to resign your position as the commander of the Eighth Army then go to **Decision 183**. If you think that something can yet be salvaged from all of this then go to **Decision 137**.

180

Your two armoured divisions, supported by truck-borne infantry, fan out to the west. By midday the lead elements of the 10th Armoured Division are closing in on Tel el Aqqaqir, on the track that runs to Sidi Abd el Rahman on the coastal road. So far no serious opposition has been encountered.

Lumsden is concerned that significant enemy forces are on his flanks and because he has not located the bulk of the enemy armour yet, he risks being cut off, or at least attacked on the flanks. He is also concerned that refuelling could be a problem. If you want Lumsden to stop and hold the track and wait for reinforcements then go to **Decision 181**. If you want him to pull back then go to **Decision 10**. If you want him to swing north along the track then go to **Decision 54**.

181

1600 hours 25 October 1942

Reports from Lumsden seem to imply that the enemy is being forced to react to your cutting off their main supply route. Already the 1st Armoured, holding the northern part of the salient, is reporting that increasing numbers of tanks from the German 15th PZ Division are being committed. So far no major attack has developed. To the front and south of the salient, elements of the Italian Littorio armoured division are moving to block any westward advances by your units. It is becoming clear that a major engagement will be fought soon.

If this is what you want to happen then go to **Decision 101**. If you think that it is wiser to pull back to the defensive lines to the west of Miteiriya Ridge then go to **Decision 10**.

182

1900 hours 25 October 1942

You move the Greek Brigade all the way to the southern end of your line. They will cover the Pavia Division whilst your 44th Infantry Division, supported by the Free French, punch a hole in the line of Folgore Division's defensive positions, allowing the 7th Armoured Division to break through into open country.

Reports suggest that some distance behind the Folgore Division is the Italian armoured force of the Ariete Division.

You wait patiently until darkness. There is a considerable minefield to negotiate first, but not as deep or dangerous as the ones in the north. You will hold back the 7th Armoured until a breach has been made in the minefield. Then they can be released to cover the infantry and engage the enemy armour. Now go to **Decision 16**.

183

You sit at your desk in the map truck for the last time and compose your resignation. Undoubtedly the job of taking over in the short term will fall to Auchinleck. You state very clearly in your letter the shortcomings of your commanders, particularly Lumsden and the other armoured commanders. However you fall short of blaming them specifically for the failure of your operation. That you reserve for yourself. To have been able to reorganise and assemble such a force in such a short time has been crippling, particularly in the face of constant urging to go on the offensive from London and Churchill in particular. Perhaps you should have resisted their calls for a little longer and given yourself more time? As you seal the envelope and call a courier to take the letter to Auchinleck, you know that your command is over.

184

1500 hours 26 October 1942

The battle in the north continues as one of a war of attrition. You decide that you will thin out your front line and create a reserve that can be used to exploit a break through. All of this will need time. It also seems that the enemy has committed his remaining armoured units to the north.

Throughout the day your air force has been bombing units of the 21st PZ Division and the Ariete Division, who are moving up from the south. This reinforces your decision to focus on the north. If the enemy can be broken here then there will be nothing left to prevent you from sweeping him off the battlefield and routing the enemy.

You focus on pulling back the New Zealand Division and the 10th Armoured, and you decide to move up the 7th Armoured from the south. Now go to **Decision 153**.

185

0200 hours 26 October 1942

You deploy your strongest force, the 44th Infantry Division, against the southernmost tip of the enemy front. The Free French will attempt to break through on the line slightly further north.

Reports suggest that lead elements of the 44th have broken through the minefield and are engaging Italian infantry around the strongpoint at Qaret el Himeimat. They have run into Italian reserves from the Pavia Division that were positioned on the El Taqa Plateau. It seems that the Italian commander is feeding them into the fight to try and secure the breach.

If you want to unleash the 7th Armoured Division now then go to **Decision 145**. If you feel it better to wait until the 44th has gained it objectives then go to **Decision 186**.

186

0600 hours 26 October 1942

The 44th Infantry have firmly established themselves along the front line of the enemy defensive positions. The Italian counterattacks from the El

Taqa Plateau have slackened off now there is daylight. Your air force is active and picking off transports and troop concentrations. It does appear that the Italians are beginning to fall back. You suspect they will try and re-establish a line hinged on Jebel Kalakh, but this will also mean that the Folgore Division will have to abandon its defensive positions or face being cut off.

Now might be the time to commit more forces to the south? If you want to shift one of your armoured division in the north down to the southern break through then go to **Decision 187**. If you feel that your forces in the south are sufficient for now then go to **Decision 216**.

187

1200 hours 26 October 1942

You convene a conference with your senior commanders to discuss the situation. Time is pressing and you need to make a decision about reinforcing the south or sticking to your original plan and focusing on the north.

The commanders are split in their opinions. Shifting units now will not only expend valuable resources, but could jeopardise the whole operation. With the situation static in the north but fluid in the south it is ultimately your decision.

If you wish to shift one of the armoured divisions to the south then go to **Decision 147**. The alternative is to call off the southern offensive and refocus on operations in the north. If you wish to do this, go to **Decision 188.**

188

1400 hours 26 October 1942

You flash off a signal to Horrocks to tell him to pull back from his advanced positions and re-establish his defensive line. Almost immediately he is on the telephone to argue his case.

'We have an opportunity here, don't throw it away,' he tells you.

'I applaud your aggressiveness. I could do with that approach here in the north. Nonetheless the bulk of our forces are here and our primary aim must be to cut the coastal road and force the enemy back. He can be broken here.

I fear that the southern attack will lead to a simple realignment of his lines,' you explain.

Reluctantly he agrees with you.

'Rest assured when the time comes you will have your chance,' you reassure him.

Now turn to **Decision 106.**

189

1800 hours 26 October 1942

It amazes you that the enemy has not made a really concerted attempt to engage your forces. It has played into your hands; being able to deal with them in a war of attrition, with the odds in your favour.

According to Horrocks the enemy has now fallen back out of artillery range. His left flank is completely uncontested. To the left, or the west, is a track that heads northeast behind Jebel Kalakh and then into the heart of the enemy supply routes and rear. Two clear options are open to you. Either Horrocks can strike west and pick up the track and use it to threaten the entire flank and rear of the enemy army, if so go to **Decision 235.** Alternatively, the 7th Armoured can cover the infantry in their crumbling operations to roll up the enemy defensive line. If you wish to do this go to **Decision 21.**

190

0600 hours 27 October 1942

Your new line stretches from the Munassib Depression in the east to the El Taqa Plateau in the west. It does look like a new enemy line is being set up using the Jebel Kalakh feature as a pivot. There have been some attempts by enemy armour to probe your new line, but nothing has come of these incursions. What still remains is the fact that the bulk of the enemy army is in the north, with relatively secure supply routes, but his flank is dangerously exposed. If you want to wait for developments here in the south then go to **Decision 150.** If you want to order an attack by your two armoured divisions in the south then go to **Decision 149.**

191

0600 hours 28 October 1942

A major thrust and attack from the south is something that you had never seriously considered, but it is now becoming clear that with the stalemate in the north the options to you in the south are considerable. You need to achieve this by taking more units out of the line in the north, whilst not compromising your overall position. If you feel that this is worth the gamble, then go to **Decision 22**. If you feel that you need instead to refocus on issues in the north and it is there that the major push should be made then go to **Decision 219**.

192

1600 hours 26 October 1942

Hastily you contact your battery commanders; they have already outlined a fire plan to deal with this sort of situation.

Moments later you begin to hear the whine of shells and the less distinct sounds of the explosions in the distance. As the minutes pass the shelling reaches a peak and remains at that level for some minutes.

'Reports?' you demand.

Your staff scurries around, trying to make sense of the fragmentary information trickling in from the front.

'One report suggests they've closed to within 500 yards'.

'But this one says they're pulling back!'

This could well be the opportunity you have been waiting for; drawing the enemy armour out and destroying Rommel's only mobile reserves. If you want to continue the bombardment, go to **Decision 74**. If you wish to order a ceasefire and examine the situation then go to **Decision 117**.

193

1200 hours 26 October 1942

The 1st Armoured Division is in grave danger of being cut off and destroyed. The 10th Armoured are under enormous pressure from the enemy tanks and anti-tank guns.

You need to act decisively; otherwise everything that you have risked and now gained will be lost. It is clear to you that the 1st Armoured may need to pull back. If you wish them to do this go to **Decision 23**. Alternatively, you need to move the 10th Armoured to support the 1st and make a concerted attempt to bring Rommel's armour to battle and destroy them. If you prefer this course of action then turn to **Decision 67**.

194

0800 hours 28 October 1942

Rommel has committed his 21st PZ Division to retake the Snipe position and waves of German and Italian tanks are being thrown at the Rifle Brigade and their supporting tanks.

Luckily the Rifle Brigade has a dozen anti-tank guns. The attacks are beaten off, leaving nearly thirty enemy tanks destroyed around the position. As the Germans and Italians withdraw from the position you order your air force and artillery to do everything to stop them from forming up again.

With the enemy forced back and seemingly unwilling to try to attack again your attention shifts to the north. The Snipe and Woodcock sector should remain on the defensive and the Australians should try to break through via Point 29 to the coast. Now go to **Decision 113**.

195

0800 hours 29 October 1942

Against all odds the Australians managed to reorganise in the darkness and press on. Two hours ago they reached the road and the railway and are sitting astride both of them, desperately trying to dig in as the sun rises.

The position for the Axis troops in the salient is precarious, but so too is that of the Australians. You realise that this situation suits you more than Rommel. He must be thinking that your next major push will be in exactly the same sector. You determine to keep him guessing, whilst you build up your reserves and consider your position. You know that Rommel will have to commit men and tanks to try and dislodge the Australians and their armoured support. For now he must react to you, while you plan a decisive blow elsewhere. Now go to **Decision 71**.

196

2400 hours 31 October 1942

You had hoped to precede the attack with a determined aerial assault and then a solid artillery barrage. Time, however, has not been on your side. A few sorties have been flown, only serving to alert the enemy as to your intentions. The artillery begins opening fire, covering the advance of the lead units.

The 151st Brigade from the 50th Division and the 152nd Brigade from the 51st, together with the 133rd Royal Sussex Infantry Brigade, will aim to advance through the minefield and clear a path to allow the 8th and 50th Royal Tank Regiments to pass through. Once the troops have achieved a breakthrough to the Rahman Track the 9th Armoured Brigade will break through, followed by the 1st Armoured Division. They will bring Rommel's tanks to battle and destroy them.

As the infantry advance they are brought under murderous fire from the German and Italian anti-tank guns. Casualties are mounting and very little progress has been made so far.

If you want to call off the attack then go to **Decision 116**. If you want to press on then go to **Decision 158**.

197

0100 hours 2 November 1942

Operation Supercharge is underway. The two main assault infantry brigades have taken their key objectives with the minimum of losses.

On the right of the main attack the Maori Battalion has also moved forward and on the left the 133rd Lorried Infantry Brigade has made good ground.

So far the 9th Armoured Brigade, due to start its breakthrough attempt at 0545, has struggled to get forward. They are due to attack towards Tel el Aqqaqir, behind the cover of a barrage. According to reports only 94 of their 130 tanks are ready for action. Their readiness is critical to the effort.

If you are prepared to postpone the next phase of the operation for half an hour to give them time to reorganise then go to **Decision 30**. If you feel that the whole operation has failed due to the 9th Armoured Brigade and want to call it off then you need to contact Churchill with the bad news and go to **Decision 234**.

198

1900 hours 26 October 1942

Your offensive is now well ahead of schedule. The Australians report that they have seized the coastal strip and are in firm control of Sidi Abd el Rahman. The German 164th and the Italian Bersaglieri have either been pushed back further west, or they are surrounded in ever-decreasing pockets of resistance.

Rommel has been throwing the remnants of the 15th and 21st PZ Divisions at your armoured units. The scattered units of the Italian Littorio Division are in less evidence than before, having suffered ruinous casualties and are clearly running out of fuel.

If you wish to continue the offensive purely in the north then go to **Decision 31**. If you feel you can risk an all out offensive along the line then go to **Decision 75**.

199

2100 hours 27 October 1942

As Rommel's remaining forces slip away to the west you pause before ordering a pursuit.

'Gentlemen, this is a complete and absolute victory,' you announce at a press conference.

Eventually you will retake Tobruk and even Benghazi, but Rommel's key formations, so resilient and dogged in attack and defence, have survived to fight another day.

Your men must chase them across the coast of North Africa until you can bring them to battle once again. This time there can be no mistakes or hesitation. Several more weeks of desert warfare remain until you face Rommel once again on the Tunisian frontier.

200

0800 hours 27 October 1942

The bad news is that the 10th Armoured has made no progress through the night. In fact the resistance in their sector seems to be stiffening. Something needs to be done in order to keep the offensive moving and you need to carefully consider your options.

In the coastal strip the 1st Armoured is still reorganising, but the 9th Armoured has yet to really come to grips with the enemy. The 90th Light Division facing them is content to hold their line and take the odd shot against your tanks should they venture too close.

If you want to shift the 7th Armoured to support the 10th Armoured and move the 1st Armoured back into line then go to **Decision 224**. This will allow you to try to push forward to take the Rahman Track and perhaps the Telegraph track. If you want to order the 7th Armoured to attack the 90th Light Division then go to **Decision 77**.

201

1230 hours 2 November 1942

Before the 1st Armoured Division can deploy properly and make their numbers count the German and Italian tanks have rolled up your front line. The whole situation is chaotic; your Lorried Infantry is slaughtered and the loss in materials is ruinous.

Finally the 1st Armoured begins to engage, but in the midst of shattered equipment and milling crowds of infantry hunting for cover it is difficult for them to make their numbers tell. It takes several hours to stabilise the situation. Rommel has delivered one of his trademark counter strokes and has also managed to extricate the bulk of his armour, pulling back to the Rahman Track and digging in. This battle is by no means over.

It is not long before Churchill calls you to demand an explanation. Clearly he has been watching the situation very closely. Now go to **Decision 234**.

202

1900 hours 3 November 1942

You order the 151st to proceed to their allotted objective. Aerial reconnaissance suggests that the intended objective line has actually been abandoned by the enemy. As the 151st moves forward it comes under heavy and accurate fire from hidden anti-tank guns and mortars. Undeterred, the 151st presses on and reaches the Rahman Track, only to see a new line of enemy resistance some distance to the west. If you want them to hold in their positions on the track then go to **Decision 81**. If you think they should push the enemy further west then go to **Decision 124**.

203

0800 hours 4 November 1942

It has taken a whole day to bring about the destruction of the Ariete Division. Along with it the majority of the Littorio and the Trieste Divisions have also been shattered. In many cases the Italians literally fought to the last man and the last round. As your troops advanced the initial resistance was impressive, but your tanks did the majority of the damage, with burned out Italian armour and vehicles scattered over a wide area of the desert.

By late morning you have created a twelve-mile hole in the enemy front. Strong columns of your tanks and Lorried Infantry are flooding west. You receive word that what remains of the Ariete and Trento Divisions have been encircled. You receive word that the enemy is trying to fall back to Fuka, some distance west of the Rahman Track. Now go to **Decision 39**.

204

0300 hours 4 November 1942

The 2nd Armoured Brigade clatters forward into action, covering the flanks of your deployed infantry. A light barrage from German artillery peppers your positions. Immediately your own artillery responds, sending shells into the darkness towards the supposed positions of the enemy guns. The

German tanks close to within 500 yards; your anti-tank guns open fire first, shattering the lead enemy vehicles. As the second wave of German armour tries to work its way around the burning tanks your own armour opens fire. The Germans, only able to target the muzzle flashes, fire back with limited results. Realising that the attack has failed, the German tanks begin to pull back. Your armour advances beyond the anti-tank line, firing at the reversing enemy. Now go to **Decision 165**.

205

0800 hours 6 November 1942

The 1st Armoured is determined to make up for lost time, having been moving all night. In the darkness the armour became separated from their support vehicles. As a consequence they have run out of fuel. The reports suggest that the fuel was exhausted around dawn. The lead elements of the division are still sixteen miles short of Bir Khalda. This unfortunate situation is only going to give Rommel more time to get away, or at least to set up a defensive position ahead. It is imperative that the 1st Armoured Division gets to Mersa Matruh as soon as possible. If you want to focus on the support trucks getting to them then go to **Decision 207**. If you want to order them to halt and wait for full fuel supplies then go to **Decision 229**.

206

1200 hours 5 November 1942

A vague memory of poring over maps of the area makes you think about the minefield. You must have discounted it for some reason otherwise you would not have chosen this route of approach for the New Zealanders. It could be that the minefield is a dummy one and that there is no danger at all. Alternatively, it could be a fully sown minefield and would cause huge casualties, leaving the New Zealanders exposed to the enemy as they work their way through it. If you want to risk it being a dummy minefield and tell the New Zealanders to press on, go to **Decision 41**. If you think it better to tell the New Zealanders to work their way around it then go to **Decision 128**.

207

1100 hours 6 November 1942

The forward supply trucks have linked up with the 1st Armoured Division at El Daba. The 7th Armoured, close to Sidi Haneish, has run into elements of a German reconnaissance group, backed up by the remnants of the 21st PZ Division. The divisional commander is keen to engage as soon as his tanks have been refuelled. It appears that the Germans are somewhat southwest of Sidi Haneish.

Heavy rain has now begun to fall, slowing the whole process of resupply. The convoys of trucks moving along the dirt tracks are reporting that in places the tracks are already impassable and they are being forced to make detours. Once the 7th is refuelled do you want to order them to engage? If so go to **Decision 230**. If you think they should flank the Germans and press on westwards then go to **Decision 43**.

208

1400 hours 5 November 1942

The 10th Armoured Division has burst through small pockets of enemy troops. You can offer them little in the way of support from the air as the pilots are unsure about targets. It becomes clear from radio intercepts that Rommel had intended to set up a new line of defence at Sidi Barrani, some eighty miles to the west of Mersa Matruh. The idea was that if he could hold this narrow passage it would give time for the bulk of his slower moving troops to retreat through the passes at Halfya and Sollum.

Your sudden move of sending the 10th Armoured forward has forced him to abandon Mersa Matruh at least a day early. You now need to press on and force the passes at Sollum and Halfya as soon as you can. Now go to **Decision 88**.

209

1800 hours 7 November 1942

You fly up to Sidi Barrani to personally congratulate your lead units. As you touch down and the light aircraft taxies along the dirt runway you can see a delegation of officers surrounding a number of Germans. There seems to be great excitement as you walk over to them and soon you realise why.

'General, may I present Field Marshal Rommel,' an excited captain from an armoured car troop announces. 'We found him five miles to the west of here, about thirty minutes ago. As soon as we realised you were coming we rushed him here to meet you.'

Your victory is complete. Rommel has been captured and his army destroyed. The war in North Africa is not quite over, but perhaps this is the beginning of the end for the Axis.

210

Reluctantly you give the commander permission to fall back behind the ridge. You feel that all elements of surprise have been lost. You continue to ponder your options. Perhaps a determined effort can be made if all of the artillery available is used to put smoke down on the other side of the ridge to cover the advance. You quickly send a command to Lumsden, commanding X Corps, which consists of the two tank divisions:

> 'It is imperative that your two armoured divisions get on through XXX Corp's objectives before the enemy can re-establish their lines. The 1st Armoured Division is being held up by enemy forces containing the end of the northern corridor. I will arrange fire support to get the 10th Armoured Division through the New Zealand front. We will use all available artillery support, if possible I want the operation to be launched this afternoon.'

Now turn to **Decision 46**.

211

'What news of operations in the south?'

The target of the 7th Armoured Division and the supporting Free French Brigade had been to take Qaret el Himeimat as their first objective. The main point of the attacks was to draw off enemy armour to the south. If the attacks were successful and broke through the enemy line then the units were to attack enemy forces engaged against X Corps from the rear. Initially at least the Allied forces in the south would face two Italian formations; the 17th Pavia Infantry Division and the paratroopers of the 185th Folgore Division. Supporting their flank was the German paratroopers of Ramcke's Brigade. In reserve to the Italian and German rear were the armour of the 21st PZ Division and the 132nd Ariete Division. News is good. The 7th Armoured Division has broken through and is believed to be in open country, beyond the rear of the Folgore Division. Shifting resources to the south will take time. If you are happy with their progress and feel that your attention should be on the north again then go to **Decision 133**. If you do want to focus on the south now then go to **Decision 175**.

212

1500 hours 24 October 1942

You receive a report that confidently tells you that the forward armoured brigade of the 1st Armoured Division has found a gap in the enemy front and will go through it as soon as mopping up operations have been completed. You contact Lumsden at X Corps HQ for confirmation.

'The situation is unclear at the moment,' he admits. 'Besides, I'll not be content until the division reaches the depression.'

The depression is Kidney Ridge, on the northern extremity of the 1st Armoured's objective.

'What about the southern corridor? What's happening down there?' you demand.

'I understand that the 8th Armoured Brigade has begun a reconnaissance attack across the ridge, or at least that was the plan,' he tells you somewhat vaguely.

'I want to know exactly what is happening as soon as you have news,' you shout.

Now turn to **Decision 233**.

213

'Prime Minister, I assure you that the plan is sound. Agreed that the armoured operations are behind schedule, but our infantry have taken all their objectives. The initiative is still with us,' you begin.

Churchill seems calmer now, but he tells you how important a victory is for the war effort, for the people of Britain who have faced such trials and tribulations with such fortitude. He goes on to tell you that should any commander fall short of expectations then you have his backing in removing them.

'I understand Prime Minister. With hope we will have news of a victory soon,' you respond.

The stakes are high and the Prime Minister's expectations are clear. Quickly you fire off a string of orders to Lumsden and the other commanders. The two armoured divisions must be out in front of the infantry by dawn. There can be no further excuses. Exhausted, you decide to take a few hours sleep while you can. Tomorrow will be as challenging as today. Now turn to **Decision 176**.

214

2300 hours 25 October 1942

The approaches to Point 29 seem to be promising, particularly after capturing a prisoner with a sketch map of the area, which revealed the paths through the minefields.

The Australian infantry creep forward to attack the other defensive perimeter around the enemy strongpoint. These are swiftly overcome, but as the reserve units move up to support the leading companies they come under tremendous fire from the main German position. Well hidden machine gun posts pin down the Australians. Any gains that have been made are lost as the Australians are forced to pull back. Nonetheless, some ground has been gained by dawn. It is clear that the Germans intend to hold on to Point 29 for as long as is possible. Now go to **Decision 55**.

215

0800 hours 26 October 1942

The great news is that Lumsden's 1st Armoured Division has taken Sidi Abd el Rahman. It has taken up defensive positions around the town as the Australians move in to occupy it. The 10th Armoured Division has been in action all night, first fending off attacks by the German 15th PZ Division and then by elements of the 21st PZ Division. Rommel has moved up his last German armoured reserves to thwart your attempts to crush him in the north.

If you want to reinforce your northern effort by moving up the 7th Armoured from the south then go to **Decision 110**. If you think that you have sufficient focus in the north and may need the 7th in the south at some point then go to **Decision 152**.

216

1000 hours 26 October 1942

The Pavia Division and supporting units from the Ariete Division have failed to make any real impression on your gains in the south. In fact it seems that they are pulling back. There is also some evidence to suggest that the Folgore Division to the north of the Pavia Division is trying to realign their flank. Of greater concern is the fact that spotter planes have seen the 21st PZ Division massing around the Jebel Kalakh feature to the northwest. It does appear that they are preparing to attack.

If you want to order the 7th Armoured Division to move and engage them before they are ready then go to **Decision 62**. If you would prefer to order your tanks to dig in and prepare for them then go to **Decision 105**.

217

1500 hours 26 October 1942

Horrocks is out in front with his main force of armour and is able to give you a commentary of the engagement.

'They are coming on slowly. I can make them out in the haze,' he tells you. 'I'm certain he's not committing his full force against us'.

'Good, you can pick them off in detail,' you reply.

The minutes pass and you can barely hear what is going on from the booming sound of the field guns and the anti-tank fire.

'The attack is breaking down!' Horrocks yells over the din. 'Yes, there are burning tanks and abandoned vehicles all across the front!'

'What about our casualties?' you demand.

'Surprisingly light; less than five per cent I'd say,' he replies. 'What do you want me to do?'

If you want to order him to stand and fight then go to **Decision 189.**

If you want to order him to counterattack then go to **Decision 20.**

218

1000 hours 27 October 1942

Horrocks reports that his 7th Armoured Division has reached the main Axis supply route that runs northeast towards the coast. The 7th have brushed aside a major attempt by the 21st PZ and the Ariete Division to stop them.

The 10th Armoured, working with the 44th Infantry Division, have pushed up north and are engaged against the Breschia Division. There is now an opportunity to finish the enemy but you will need at least twenty-four hours to reorganise for the last major push. Now go to **Decision 191**.

219

1800 hours 27 October 1942

In the north the enemy has focused on throwing armour at your defensive positions around the Kidney Ridge area. You determine to push the enemy out of two positions near Kidney Ridge that have been troublesome for your men. To the northwest is a feature codenamed Woodcock, to the southwest another feature codenamed Snipe. It will fall to the 1st Armoured Division to plan assaults on these positions tonight. If you want to order that both features be attacked then go to **Decision 112**. If you think that one is sufficient then go to **Decision 154**.

220

1600 hours 26 October 1942

The German and Italian tanks close to within anti-tank gun range, firing at your positions as they clatter forward. You listen intently to the radio traffic as the battle intensifies. For Rommel actions like this are dangerous, with his lack of resources, but you know that he has won victories against all odds before. Your men need to hold, to knock out as many of the enemy tanks as possible and to bleed him dry of reserves.

The enemy tanks are now within 200 yards of your defences. The anti-tank guns have taken a fearful pounding and your tanks have suffered too. But reports suggest that several dozen enemy vehicles are blazing in the heat of the afternoon. You now need to decide how to respond to this counterattack and go to **Decision 159**.

221

1200 hours 26 October 1942

Lumsden paints a pretty bleak picture of the situation. He tells you that both of the armoured divisions are on the point of destruction. You know that losing these units will ruin any chance you have of being able to exploit your advantageous position.

'The 1st Armoured need to be pulled out,' he tells you. 'They risk being overrun at any time'.

You wonder if Lumsden is overreacting or whether he is secretly hoping that by his actions and advice you will be discredited. There is obviously no love lost as far as he is concerned. If you want to agree with him and pull back the 1st Armoured Division then go to **Decision 23**. If you feel it better to reinforce them with the 10th Armoured and push on then go to **Decision 67**.

222

0800 hours 1 November 1942

You decide to call off the attack. To lose all of your carefully assembled reserves would mean that any chance of dislodging Rommel will be lost.

Equally, you realise that with all of the assets at your disposal you need to ensure that every advantage is used. This means making full use of your air superiority and artillery.

You determine to launch the offensive slightly further south. Mercifully the attack you have just called off has not caused too much damage to your units. You order them to return to reserve positions and decide to launch the attack tonight. Throughout the day your aircraft will fly as many sorties over the enemy front as possible and for hours before your lead units get underway your artillery will pound the enemy positions. Now go to **Decision 197**.

223

0600 hours 2 November 1942

It is with a heavy heart that you call off the attack and order the men to fall back to their start positions. You know that Churchill will never forgive you for this lack of spirit and determination. You take one last look at the map board and think about what may have been. You then make that difficult call to Churchill and go to **Decision 234**.

224

1200 hours 27 October 1942

An hour ago the lead elements of the 7th Armoured Division crossed the Rahman Track. Overhead your aircraft have been strafing enemy transport all along the track and paralysing Rommel's resupply efforts. Already your commanders can see the fruits of their labours; abandoned German tanks and vehicles outnumber those knocked out in combat. Rommel's fuel problems were obviously far more critical than you had supposed. There is still opposition on parts of the Rahman Track, where the remnants of the German armour are fighting for their lives. Now turn to **Decision 118**.

225

1100 hours 2 November 1942

Dodging your artillery bombardment as best they can, the 15th, 21st and Littorio Armoured Divisions advance on your positions. Your commanders report several of the tanks being abandoned as their tracks are thrown or concussive blasts neutralise the crews. The tanks keep on advancing and the anti-tank guns now engage the enemy. Rommel has waited for this moment to pinpoint the positions of your batteries. He now directs his artillery straight onto your anti-tank positions.

It is a well-timed and perfectly executed attempt to knock out your key defensive line. Casualties are high amongst the gun crews and their fire slackens off as the hits begin to tell and the tanks close in. If you feel confident that they can hold then go to **Decision 79**. If you want to order the 1st Armoured to mount a counterattack now then go to **Decision 122**.

226

0115 hours 3 November 1942

As far as your commanders are concerned Rommel cannot have more than fifty operational tanks left. Added to this, the gun screen near the Rahman Track is very thin and only needs one more concerted effort to break through it. The 7th Armoured Division will attempt to break through a two-mile front to the north of Tel el Aqqaqir.

You need to quickly decide whether you will not only risk the deployment of the 7th Armoured, but also whether you will use the 2nd and 8th Armoured Brigades to achieve the necessary gap first. If you think the 7th Armoured is sufficient for the task then go to **Decision 80**. If you think it prudent to use the 2nd and 8th Armoured Brigades as well then go to **Decision 123**.

227

0730 5 November 1942

The 1st Armoured Division swings into El Daba to find that Rommel's rearguard has gone. The 10th Armoured Division, now leading the race to Fuka, is still short of their target. The New Zealanders report that they are held up on the outer perimeter of a minefield. They have come across a wire fence stretching out across the desert and beyond it they believe that the Germans have dug in.

If you want to send a reconnaissance aircraft to check then go to **Decision 168**. If you think that the minefield is familiar then go to **Decision 206**.

228

0800 hours 5 November 1942

The 1st Armoured Division has halted as ordered and has been awaiting fuel resupply. It arrives just after first light. Your supply columns have worked miracles; driving their dangerous loads through the night on uncleared roads, with pockets of enemy troops still at large in the area. The division reports that they will be ready to get underway at around 1100. The question is whether it is too late to try and catch Rommel's rearguard. If you think the drive for Mersa Matruh is still important then go to **Decision 239**. If you want to call off the pursuit then go to **Decision 42**.

229

1200 hours 6 November 1942

It has started to rain. Already the dirt tracks are breaking down and the trucks are finding it increasingly difficult to make progress. With the lead armoured units stationary and awaiting supply you can feel Rommel slipping through your fingers. Each passing hour gives him time to extricate more of his men and equipment to the relative safety of the west.

The 10th Armoured Division, on the coastal road, is awaiting orders. They have been making for Fuka, but you wonder whether it is time for them to take the lead in the pursuit. If you want to make them the lead chasing division then go to **Decision 87**. If you want to call off the pursuit then go to **Decision 42**.

230

1400 hours 6 November 1942

In what is destined to be the last pitched battle for the Egyptian border region, the 7th Armoured swings into action against the 21st PZ Division and supporting German troops.

The German commander, aware of the danger he is in, fights a careful battle, conscious of the fact that he is short on fuel, ammunition and resources. He has clearly ordered his tank crews to only take necessary risks.

Each time the 7th Armoured attempts to flank or encircle the enemy the enemy simply fall back to a new position. It is a battle the Germans cannot win. By the end of the day the 21st PZ Division has lost at least sixteen tanks and most of their anti-tank guns, with their experienced crews either killed or captured.

The action has cost you valuable time. Rommel gambled on the ability of the 21st PZ Division to hold you up long enough to let more of his troops slip to the west. Now go to **Decision 131**.

231

'Get them forward, no excuses. Tell them that their armoured shield is essential to protect the infantry. The infantry cannot be expected to hold the front and start reducing the pockets of enemy resistance too!' you tell your staff officer.

Staring at the map, you begin to realise that the whole operation hangs in the balance. With insufficient dash and momentum the entire plan could fall apart. If the enemy can contain you because of your troops' inactivity then you are back to square one. Surprise will be lost and the enemy will be able to shuffle his forces and re-establish a new line. Next time it will be even more difficult. Time is not on your side; Churchill needs a victory and has given you all the tools to achieve one. This may be your only chance.

Now turn to **Decision 133**.

232

'Lumsden, let me speak plainly to you if I may. You must drive your division commanders. If they hang back any more then I will have no alternative but to replace them. Is that clear?'

'Perfectly clear,' he replies. 'We're planning to break through in the north by the afternoon and in the south we're mounting a reconnaissance in force across the ridge.'

'I want your armour out front in their positions. That means Kidney Ridge in the north and a solid force in the south,' you confirm.

'Understood. I'll keep you appraised of progress,' he replies.

Now turn to **Decision 212**.

233

1845 24 October 1942

Finally a report comes in from Lumsden. Quickly you devour the news, shaking your head as you read it:

'The 8th Armoured Brigade is in hulled down positions on the northern side of the Miteiriya Ridge. They have run into minefields that have been freshly laid. The minefields are covered by anti-tank guns, machineguns and tanks. We believe the latter to belong to 21st PZ. Elements of 8th and 9th Armoured Brigades have been engaged in contact with the Italian Littorio Division. We have suffered two losses and believe we have knocked out seventeen enemy tanks. We are planning to launch another operation tonight to get the bulk of the 10th Armoured Division out into the open. Presently I am unaware of the location of the 24th Armoured Brigade.'

This may all be very well, but you have no news of the 1st Armoured Division. If you want to demand that information then go to **Decision 92**. If you feel all that could be done has been done then go to **Decision 135**.

234

'I cannot guarantee you a victory yet. We must wait for news,' you tell the Prime Minister.

You can tell that this is not what he wants to hear. He rants and raves for a few moments, telling you how he had to put up with incompetent senior commanders when he was in the Malakand and in the Sudan. There was worse in South Africa. How could such a great empire like Britain have been reduced to mediocrity? Patiently you listen to him then apologise. It is the worst thing to do, as it only allows him to begin again by telling you how much it has cost him to get the men and resources for your army.

'I'm planning the operation under the direct command of Alexander,' he tells you.

Stunned you can barely believe what you are hearing.

'Return to England as soon as you can,' he barks at you. 'You are relieved of command.'

Your command of the Eighth Army is over.

235

0600 hours 27 October 1942

You take the decision to shift the 10th Armoured Division south overnight and throw it straight into the attack. It can cover the operations of the 44th, leaving the 7th Armoured to strike out and make inroads into the flank of the enemy. It was a difficult decision and fraught with complications, but you hope it will be worth all of the problems it has posed. Now turn to **Decision 149.**

236

1100 hours 2 November 1942

Your artillery bombardment has shifted to the concentration points of the enemy armour. Through their field glasses your front line commanders can see several of the tanks burning in the distance. Rather belatedly, the Germans resort to counter battery fire, trying to hit your own artillery concentrations in the rear. Whilst they try to do this, firing virtually blind, your front line troops and tanks are safe and are able to add their long-range anti-tank fire against the handful of enemy tanks that have crept forward.

Soon it is all over; your commanders can see that the remnants of the enemy armour have pulled back with unsustainable losses. Your own casualties are remarkably light. You now feel that by enlarging the gap and providing the optimum conditions for the 7th Armoured to break through into open countryside is your key priority. You hope that they will be able to pass through your front line and swing north to capture the railway station at Ghazal. Now go to **Decision 163**.

237

2200 hours 4 November 1942

The New Zealanders have reached Fuka ahead of the rearguard of Rommel's army. It seems that the remnants of the 15th PZ Division are still falling back towards Fuka. The New Zealanders, having successfully negotiated a dummy minefield, are now preparing to dig in at Fuka and hold the coast road to prevent the 15th PZ Division from slipping past. You need to act quickly and decide how to respond to this opportunity. If you want the New Zealanders to dig in and wait for the 15th PZ Division and then hold them until your armour can catch up, go to **Decision 84**. If you think they should push on up the coast towards Mersa Matruh and leave the 15th PZ Division to your armour then go to **Decision 127**.

238

1200 hours 5 November 1942

For what seems like an eternity you wait for news that the fuel tanks have caught up with your chasing divisions. Finally the reports begin to filter in that the 1st and 7th Armoured Divisions should be ready to move again within the hour. The support vehicles are still being refuelled and will follow on as soon as possible. It is now possible to resume the pursuit.

If you want to order the pursuit to continue, then go to **Decision 86**. If you think it is already too late to catch Rommel then go to **Decision 42**.

239

0800 hours 6 November 1942

You are at least a day ahead of schedule. So far resistance has been limited and you hope that your divisions can remain on the move. The more of Rommel's forces that you capture and destroy now, the greater the likelihood of your being able to chase him all the way along the coast.

Your attempts to snare Rommel at El Daba and Fuka have failed, but the 7th and 10th Armoured are just ten miles from Mersa Matruh. The

1st Armoured has overtaken elements of the 21st PZ Division to the east of Maaten Baggush. Having run out of ammunition and fuel the Germans were forced to surrender. The 1st Armoured has handed over the prisoners to the New Zealanders and are ready to press on to the west.

If you think you have done sufficient damage to Rommel and will call a halt to the pursuit then go to **Decision 130**. If you want to press on to Mersa Matruh then go to **Decision 171**.

240

1800 hours 7 November 1942

You fly up to Sidi Barrani to congratulate the men. It has been a difficult victory, tinged with regret that your carefully-laid plans did not go according to schedule. Yet the victory is nearly a complete one. Rommel's army has been comprehensively defeated. His German core has managed to escape, albeit depleted, and the Italians in his army have largely been captured.

You know that the war for North Africa is not yet over. Winston Churchill is highly complementary of your victory; it was one that he needed so badly. He puts everything in perspective when he says it is not the beginning of the end of the war, but it might be the end of the beginning. There is still much more fighting to do before Europe is free once again.

ROMMEL, ERWIN JOHANNES EUGEN

Rommel was born near Ulm in 1891. He joined the 124th Wurttenberg Infantry Regiment as an officer cadet in 1910. During the First World War Rommel won a number of gallantry awards, including the Iron Cross, fighting in Romania and France.

During the Interwar Years Rommel wrote several books on military training and was involved in training members of the Hitler Youth. By 1938 he had reached the rank of colonel.

In February 1940 he was appointed commander of the 7th PZ Division. In France in May 1940 he gained a reputation for his dash and valour. His division fought at Arras and captured Lille. The 7th reached Rouen on 5 June then struck for Dieppe on the coast. By the middle of June he had captured Cherbourg and was heading for Bordeaux when the armistice was signed.

Rommel's successes were rewarded by promotion and a new posting, this time as commander of the 5th Light Division (later the 21st PZ Division), bound for Libya. On 6 February 1941 the new German forces, known as the Afrika Korps, were under his command. North Africa would be his theatre of war until 1943. His exploits would earn him the nickname 'The Desert Fox'.

Rommel had inherited a desperate situation, but nonetheless he launched a limited offensive in March 1941. It was well timed, as vital British forces had been withdrawn to fight in Greece. The British panicked and ordered a withdrawal. Rommel was determined to keep up the pressure and had reached Tobruk by mid-April. The siege would last for 240 days.

By 1 November 1941 the British were ready to launch Operation Crusader to relieve Tobruk. Although it initially gained ground, the offensive faltered as a result of deft counterattacks. Fortunes would favour Rommel and he was ready to launch his own new offensive by 21 January 1942. The Allies lost Benghazi again, pulled back to Tobruk and dug in around Gazalla.

On 26 May 1942 Rommel threw his forces at Gazalla. A largely Italian force attacked head on whilst his more mobile units outflanked the Allies to the south. The battle to encircle the enemy failed but renewed attacks by Rommel forced the Allies to withdraw; this time Tobruk was taken, along with 33,000 prisoners.

Rommel was determined not to give the Allies a chance to dig in and prepare for him. He decided to make a thrust for Egypt. His forces surrounded Mersa Matruh, which surrendered on 29 June, along with 6,000 prisoners.

By 25 June Auchinleck had assumed command of the Eighth Army; a new foe for Rommel to face. Auchinleck established a line at El Alamein, protected in the south by the Quattara Depression. Despite a lack of supplies and equipment and the fact his men were exhausted after several weeks of continuous fighting, Rommel pushed east. This led to the First Battle of El Alamein, on 1 July 1942. Although Rommel mauled the Eighth Army, it held.

Rommel had lost momentum and he knew time was running out. Montgomery had replaced Auchinleck and fresh men and equipment were flooding into Egypt. Rommel decided to gamble on another offensive, which he launched on 30 August 1942.

His forces passed El Alamein and then headed north at the Alam el Halfa ridge. It was exactly as Montgomery had expected and Rommel found himself facing well prepared positions. What was worse was that his armour was running out of fuel. Rommel realised he would have to withdraw. He was back at his original start line by 5 September; his losses in tanks and trucks had been ruinous.

Rommel's health was failing after months of intense activity in a challenging environment. He left for Germany and Austria in September, but had left clear instructions to his subordinates in the event of Montgomery attacking before he returned. His defensive line was good, protected by deep minefields. His plan was to use the infantry to hold and the armour to face any threatened sector of the front.

As it was, the attack did come on 23 October; German counterattacks over the next two days ablated the armoured forces that were left. Rommel rushed back to North Africa, but despite all his best efforts his army was on the verge of collapse. He was desperate to withdraw before it was too late. Hitler, however, adamantly refused to allow it.

On 4 November, after the destruction of the Italian XX Motorised Corps, a 20km gap was opened up and Allied forces passed through it. Rommel began to retreat, but it was too late; most of his forces lacked transport and were overrun.

As a commander, Rommel was not yet finished in North Africa. In February 1943 he inflicted a bloody defeat on US forces in the Kasserine Pass, then returned to take up defensive positions against Montgomery on the Mareth Line.

Rommel's last offensive in North Africa took place on 6 March 1943, but Montgomery repulsed the assault. Three days later Rommel left North Africa. He served as a commander in Greece during the summer of 1943, but then was posted to command Army Group B in Normandy in November.

Rommel became convinced that when the Allied invasion of Europe did come it would be Normandy. Consequently he tirelessly worked to improve the scant defences. Rommel was on leave when the invasion came in June 1944, but he was back to direct the fighting around Caen against his old enemy Montgomery.

On 17 July 1944 Rommel narrowly avoided being killed when his staff car was attacked by enemy aircraft. Days later, on 20 July, Rommel was implicated in the failed attempt to assassinate Hitler. He was offered a choice in October 1944; suicide or a public trial and he chose the former.

After his failure at Alam el Halfa Rommel knew that neither time nor resources were on his side. His German and Italian troops were stretched to breaking point. North Africa, in terms of German priorities, was a backwater compared to the desperate struggle against Russia.

Rommel put his plans into action to resist an Allied offensive before he left for Germany on 23 September 1942. Rommel still hoped that the British would be forced to withdraw troops from North Africa to counter a potential threat of German forces sweeping into southern Russia and, ultimately, he hoped, the Middle East. All this relied on victory at Stalingrad. If this came to pass, his troops would have valuable time to prepare and hopefully to be reinforced.

Rommel set up two belts of deep minefields, which aimed to prevent Allied armour from breaking through. His main defensive positions were some 2km behind the second minefield. Collectively, the minefields consisted of around 500,000 mines. He knew that if the Allied armour broke out into open terrain then all was lost. His tactics were to hold the Allies behind the minefields with infantry and artillery and to use his limited armour forces to plug any gaps. The 15th PZ Division and the Lutorio Division covered the north and the 21st PZ Division and the Ariete Division the south. His armoured reserve consisted of the 90th Light Division and the Trieste Division. Rommel hoped that he could move these three forces more quickly to the crucial point on the front than the Allies. But movement was limited due to the perilous state of the fuel supplies.

If you would like to take the role of Rommel in the battle then go to **Decision 241**.

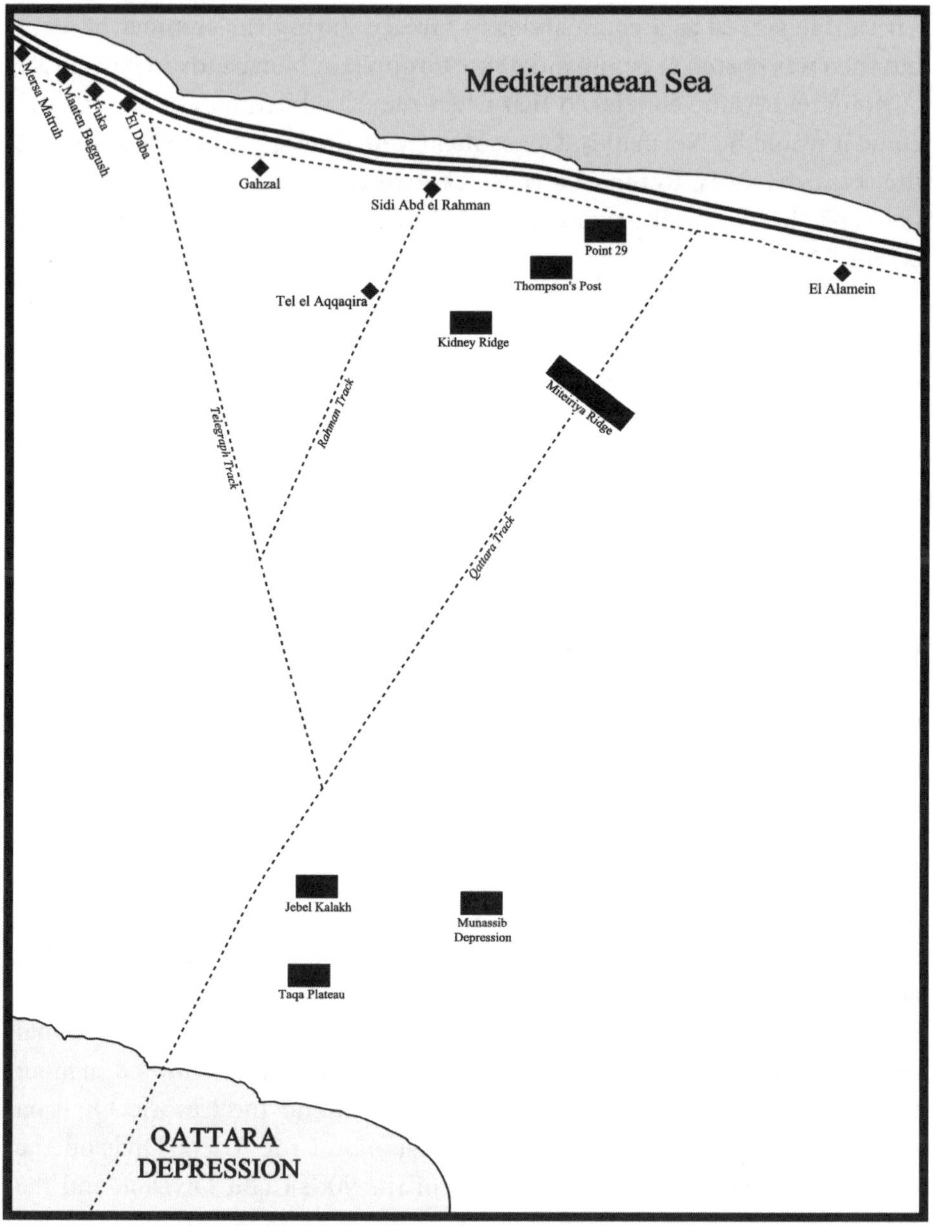

Mediterranean Sea
Mersa Matruh
Maaten Baggush
Fuka
El Daba
Gahzal
Sidi Abd el Rahman
Point 29
Thompson's Post
El Alamein
Tel el Aqqaqira
Kidney Ridge
Miteiriya Ridge
Rahman Track
Telegraph Track
Qattara Track
Jebel Kalakh
Munassib Depression
Taqa Plateau
QATTARA DEPRESSION

Battleground General: El Alamein – German

241

Victory can be so elusive. The seesaw North African campaign has exhausted you. Your health is not good and the rigours of desert warfare have caused your illnesses to become even more acute. During the summer of 1942 you gambled on the weakness of the British and beat them all the way back into Egypt. You managed to get within a few hundred miles of the Pyramids themselves. On 1 July a month of inconclusive fighting started. It was a battle of attrition and by the time you reached El Alamein you only had a handful of operational tanks.

In late August you learned that a vast convoy had reached Egypt. Auchinleck, your opposing commander, had been replaced by Montgomery. Time is running out. One last throw of the dice might have just seen you reach Cairo and seize the Suez Canal. Utilising the mobile forces under your command you opened the battle of Alam el Halfa on 30 August 1942. After passing the British El Alamein line to the south your troops were to drive north towards the Alam el Halfa ridge.

By 2 September you had realised that Montgomery had anticipated this and was dug in and prepared. You determined to pull out, realising now that the war in North Africa is unwinnable.

The British have been mounting raids against your ports and supply depots for several days. Their navies intercept the bulk of your supplies and, above all, your fuel. Less than thirty-percent of your supplies are getting through. You know the situation is perilous; you can only hope to fight a defensive war against the British. Yet you still have 115,000 men and 550 tanks, the vast majority of both being Italian. You know that the British will soon be strong enough to launch their own offensive. Your only hope is for a German breakthrough at Stalingrad. This will force the British to shift a considerable amount of men and equipment into the Middle East to forestall a German attack into Persia. All you can hope for now is that your prepared positions are strong enough.

You have created two belts of minefields, aimed to restrict or at least funnel enemy armoured attacks. Each front of your defensive boxes is thinly held; your main defensive positions are some two miles behind the second belt of minefields. You have sewn half-a-million mines. You have two relatively powerful mobile armoured groups stationed behind the defensive lines. One is in the north and the other in the south. Wherever the British thrust comes you should be able to respond, but fuel is a major problem and your armours' mobility will be limited.

Having made all your plans and leaving the army in the hands of General Georg Stumme, you leave for convalescence in Germany on 23 September 1942. As best you can you keep abreast of developments in North Africa.

On 23 October 1942 you receive an urgent order, telling you that the British have launched an offensive in North Africa and you are to return to your command as quickly as possible.

Your options are to fly via Rome - go to **Decision 266**, to fly via Athens - go to **Decision 291**, or to fly via Belgrade - go to **Decision 312**.

242

1800 hours 26 October 1942

The whole area is alive with enemy aircraft and there is a constant, rolling barrage that seeks to shatter any chance of a cohesive attack. Your men have not even closed with the enemy positions. The 15th PZ has taken irreplaceable losses and the other units are taking high casualties as each minute passes. As yet no real progress has been made. The enemy has clearly concentrated the bulk of its fire power and air support. If you wish to call off the attack before the losses become too enormous then go to **Decision 376**. If you wish to press on with the attack then go to **Decision 355**.

243

2400 hours 26 October 1942

The situation in the north is becoming critical. Your intelligence reports suggest that the enemy has massed between five and six divisions in the coastal area, two of which are full armoured divisions. Even though you have three German divisions, supported by four Italian, they are divisions in name alone. Many of them are shadow formations, depleted in equipment, men, fuel and ammunition. Your particular concern is for the Italians, as the infantry divisions lack their own transport. This makes them a liability in a mobile battle. You are sure that everything hinges on your ability to hold the line in the north. You are unsure as to whether the forces that you have deployed along the coastal strip will be sufficient. If you want to move the 21st PZ Division up from the south to support the effort then go to **Decision 244**. If you think that you are sufficiently strong already then go to **Decision 269**.

244

0800 hours 27 October 1942

The Ariete and the 21st PZ Division managed to move up under the cover of darkness and take up their positions to cover the Kidney Ridge area. It is clear to you that this sector of the front is a key one. It would only take a determined push by the enemy from this position to sever the Rahman Track and seize Tel el Aqqaqir. It also means that your most potent mobile forces are now committed to defending this fragile part of the front. Now turn to **Decision 382**.

245

2400 hours 28 October 1942

The situation is becoming disastrous and deteriorating hour by hour. Although you know that your troops in the south are probably still resisting, mentally you have had to write them off. There is very little that you can do to help them. If you risk what remains of your mobile troops in the north then you stand to lose your entire army. Your command centre has been receiving increasingly frantic messages from the south, asking permission to surrender or instructions on which routes to take and where to rally. Your forces in the north are falling back towards Fuka. Now turn to **Decision 316**

246

2400 hours 28 October 1942

The retreat has decended into chaos. It is obvious that the bulk of your troops pulling back from the southern front have been cut off or overrun. Your rearguard along the coastal strip has been under continual pressure for the past twenty four hours. The situation is so acute that it became impossible to set up any line of defence at Fuka, which is now occupied by the enemy. You are left with no choice but to continue to fall back with what scattered elements of your army that remain towards Mersa Matruh. Now turn to **Decision 295.**

247

Your troops manage to break through to Mersa Matruh just in time, before the pincers of the enemy armoured columns shut off the west. The situation at Mersa Matruh is chaotic, with scattered groups of troops from several divisions attempting to reorganise themselves and establish a defensive position. It may be impossible to do this and you may need to fall back even further to the west. If you want to hold here at Mersa Matruh then go to **Decision 272**. If you want to retreat now then go to **Decision 358**.

248

1200 hours 29 October 1942

You manage to establish a new front running from Ghazal along the coastal strip. Sidi Abd el Rahman has had to be abandoned. Your intelligence reports suggest that the enemy is massing to the south of Ghazal and that they are probing towards El Daba. Clearly they mean to cut you off from the west.

You have only two options; you can decide to try to hold here and fight by going to **Decision 273**, or to fall back to the west by turning to **Decision 316**.

249

0800 hours 27 October 1942

The Folgore Division pulls back to cover the gap between the El Taqa Plateau and the Munassib Depression. This leaves the Pavia Division somewhat exposed in the south. If you want them to hold their position, then go to **Decision 298**. If you want to order the Pavia Division to fall back to the El Taqa Plateau then go to **Decision 319**.

250

0800 hours 28 October 1942

The Free French and Greeks have caused huge damage to your supply lines and have shot up dozens of fuel trucks. Your tanks are starting to run out of fuel and ammunition. You now give them permission to begin to fall back but some have to be abandoned. All of a sudden your southern front is collapsing. If you want to try to re-establish a new line, then go to **Decision 320**. If you want to order your troops in the south to withdraw, holding the enemy as long as possible, then go to **Decision 299**.

251

The 21st Panzer Division moves into position, with the Ariete Division in reserve. Your short-lived confidence that you have stabilised the battle front is shattered by news that the enemy has launched a major assault in the Kidney Ridge area, to the north. The enemy has punched a hole in your line and elements of two armoured divisions are streaming through and have reached the Rahman Track. Now turn to **Decision 381**.

252

1200 hours 27 October 1942

The enemy is pouring troops into position around the Kidney Ridge feature. You will need to decide where you will launch your counterattack; where it can do the most damage at the least cost to your vital mobile troops. You can hope to achieve local armoured superiority, but you know that any advantage that you might have will be fleeting if the enemy can redeploy its more superior tank force. If you want to launch a counterattack against the Kidney Ridge feature then go to **Decision 277**. If you want to launch your counterattack against the northwest of the salient, against a position that the enemy has codenamed Woodcock, then go to **Decision 300**. If you wish to launch your assault against the southwest of the salient, against a position codenamed Snipe, then go to **Decision 321**.

253

Fascinated, you listen to the jumbled radio traffic between your lead armoured units and higher command. Your armour is making good progress and it appears that the enemy armour is falling back, having suffered considerable casualties. Your armoured units overrun the Snipe position. The enemy is retreating. This is precisely the outcome that you were hoping for, but you now need to decide whether or not to exploit the situation. If you wish to pull back from the Snipe position and go back on the defensive then go to **Decision 278**. If you want to continue the attack then go to **Decision 301**.

254

The enemy armour is pushing in on the flanks of your tank formation. Von Thoma believes that elements of at least two enemy armoured divisions are engaged. He is deeply concerned that your armour will be overwhelmed. If you want to order a withdrawal, then go to **Decision 386**. If you want to reinforce the attack with the 15th Panzer Division then go to **Decision 387**.

255

The enemy attacks are temporarily halted and the focus now seems to be on possession of the area surrounding the railway. In most sectors in the coastal strip the enemy advances are slowing down. You wonder why the enemy has chosen to do this, but at least it gives you breathing time to reorganise and to reappraise developments on the battlefield. If you want to leave the 90th Light Division in a forward position in case of renewed attacks then go to **Decision 324**. If you want to pull them back and put them into reserve then go to **Decision 280**.

256

As your armour begins to fall back from its advanced positions the enemy tentatively advances and retakes its positions in the salient. To your amazement they do not push forward but seem content to re-establish themselves. There is still continued bombardment all along the northern sector of the front. Enemy aircraft are in much evidence and mounting combat patrols. However, compared to the events over the past few hours, the front has gone relatively quiet. Your men take the opportunity to dig in and prepare for the inevitable renewal of the enemy offensive. Now turn to **Decision 278**.

257

The majority of the 164th and the Bersaglieri manage to evacuate the salient before sunrise. They have been forced to abandon many of their anti-tank guns and support equipment. The men have flooded back toward Sidi Abd el Rahman. It is obvious that the enemy will now attempt a similar tactic; pressure from the east coupled by an attempt to encircle from the south. Clearly Sidi Abd el Rahman needs reinforcement. The situation becomes even more grave when you begin to receive reports of enemy armour moving northwest in force, through your old forward positions and towards Sidi Abd el Rahman. It is likely from your intelligence that the enemy has committed two armoured divisions to this sector. Your choices are stark. If you wish to commit your own armour to try to counter their tank attacks then go to **Decision 325**. If you feel that you would waste your armour and that it may be needed elsewhere to cover a possible withdrawal then go to **Decision 346**.

258

You decide to continue to resist despite the mounting pressure on your troops. But disastrous news comes far earlier than you could have possibly feared. The enemy has moved into El Daba in force. There is also a rumour that Fuka has been captured. Your western flank has almost completely disintegrated and your troops are cut off from the west. You have only two options; you can try to fight your way out and go to **Decision 318**. Or you can surrender what remains of your command at Ghazal and go to **Decision 297**.

259

Your infantry in the coastal strip has taken enormous casualties from the enemy artillery and aircraft attacks. Defensive positions have been destroyed and it is almost impossible for anything to move without attracting the attention of enemy fire. Luckily your armour is safe, as it is positioned further south. The enemy barrage and focus of attack now shifts slightly south in the direction of Tel el Aqqaqir. They have set up a rolling barrage, which is being followed by strong armoured units and supported by infantry. They clearly mean to sever the Rahman Track. Your front line defence consists of infantry supported by anti-tank guns. If you feel that they are sufficiently strong to hold the enemy assault then go to **Decision 348**. If you think that you need to commit your armoured troops to assist then go to **Decision 327**.

260

As von Thoma feared, the enemy regroups and pushes through to the Rahman Track, leaving a trail of wrecked and abandoned German and Italian tanks in their wake. Very little now stands between your army and disaster; only a handful of German armour, with a limited supply of fuel. It is now obvious that unless you risk what remains of your armour in a counterattack, then your troops in the south will be cut off and lost. If you want to order the counterattack, go to **Decision 370**. If you want to order your armour to withdraw, taking with it as many mobile units as possible, and head for the northwest, then go to **Decision 392**.

261

Von Thoma does his best to assemble as many running tanks as possible to engage the enemy armour in and around Tel el Aqqaqir. He keeps you appraised of the situation and paints a picture of a confused and vicious struggle, where more tanks are being lost as they run out of fuel than to enemy action. By the time the British fall back to regroup you have barely thirty tanks left. What is more, there is no chance of being able to salvage your broken down vehicles on the battlefield. The enemy has destroyed many of your salvage trucks and even those that remain have barely enough fuel to get them off the battlefield. If you want to try to establish a line to try to hold the enemy at Tel el Aqqaqir then go to **Decision 349**. If you think that the situation in this sector has already been lost and that a general withdrawal towards the northwest is necessary then go to **Decision 285**.

262

2400 hours 2 November 1942

The shorter withdrawal to El Daba gives some of your troops retreating up the Telegraph Track the opportunity to catch up with you. Each hour sees new, depleted battalions of predominantly Italian troops, mixed with elements of Ramck's parachute brigade, reach El Daba. There are still significant numbers of Italian troops marching with grim determination northwest along the Telegraph Track. The commander of the Ariete Division bursts into your command truck. In halting German he begs you to allow him to use the remnants of his division to counterattack to the southeast and probe along the Telegraph Track as far as possible, in order to scoop up as many Italian infantry as possible. You consider his request and feel for the position he finds himself in. If you want to allow him to do this, then go to **Decision 310**. If you feel you have, with regret, to refuse him, then go to **Decision 289**.

263

The bulk of what remains of your German troops are being pushed back towards Fuka. It is becoming an increasingly desperate struggle. Each time your troops try to turn and fight they are overwhelmed. Each mile sees more of your tanks and vehicles having to be abandoned along the road as they run out of fuel. Vital and irreplaceable equipment is being lost. Further south the relative silence from your Italian formations indicates that most of them have either been captured or killed. There is considerable enemy pressure and activity all along the coastal road. Slowly but surely you are being forced back towards Mersa Matruh. You arrive there ahead of the bulk of the army, hoping to be able to reorganise and to make some sense out of the chaos. You receive a cable from Rome. The Italian High Command and Mussolini are demanding your resignation. They blame you for sacrificing tens of thousands of Italian troops. Minutes later you receive a message from Berlin; Hitler still backs you and is sure that you can turn the situation around and that ultimately victory can be Germany's. If you wish to resign then go to **Decision 309**. If you wish to stay on in command of the Afrika Korps and salvage what you can then go to **Decision 330**.

264

From previous experience you know that army high command will not give you the necessary permission for a major withdrawal. They will need to refer the matter to Hitler. Only he will permit retrograde steps. You do not know if you will have sufficient time to wait for him to make his mind up. Each hour that passes puts what remains of the army at ever greater threat. You can either choose to withdraw now and fall back towards Baggush and go to **Decision 373**, or you can choose to wait for his permission and go to **Decision 395**.

265

You have waited for too long. The enemy has cut off your line of retreat to the west. You send what you expect to be your last communication as commander of the Afrika Korps to Berlin. They instruct you to make for a dirt airfield just outside the town and use any available transport aircraft to fly west. They will arrange your onward transport to Italy. Hitler is adamant that he will not allow you to fall into enemy hands. If you wish to accede to Hitler's instructions and abandon your army then go to **Decision 399**. If you feel that it is honourable to surrender along with your army then go to **Decision 398**.

266

Your flight is relatively uneventful and you touch down in Rome, having already made plans to meet with key Italian officials. You impress upon them the importance of continuing to try to get fuel, food and ammunition through to North Africa. Patiently you listen to their excuses. It was a major mistake not to have made concerted efforts to launch an invasion of Malta. The tiny island is causing enormous problems as a base for fighter aircraft and bombers and as a hiding place for the Royal Navy, which preys on Axis shipping in the Mediterranean. They promise to redouble their efforts, but unconvinced you head back to the airfield and board a Ju52 bound for Libya. Your aircraft is escorted by a pair of Me109s. Now turn to **Decision 354**.

267

1800 hours 26 October 1942

Reluctantly you call off the attack. You know that you cannot possibly win a war of attrition against the British. Their advantages in men and equipment are just too huge for you to trade blows with them. You need to pick the time and place where you can have local superiority. You carefully examine the situation map for what it is worth, as intelligence is patchy and untrustworthy. The key decision to make is where you think the main battle will be fought. You wish that you knew exactly where Montgomery has positioned the bulk of his armour. If you could know this then you would be able to predict the likely paths through the minefields that he is hoping to use. You cannot afford to allow this armour to break through the minefields and your defensive boxes. The supply situation is precarious enough without having enemy tanks roaming about in your rear areas. If you think that the north is likely to be the main theatre of operations then go to **Decision 334**. If, on the other hand, you think that this could be a feint and that you need to continue to protect the southern part of your front then go to **Decision 313**.

268

2400 hours 26 October 1942

Over the course of the past four hours the situation has deteriorated. The enemy has poured armoured units and Lorried Infantry through the gap left by the 15th PZ Division. Your commanders on the ground report that the enemy has succeeded in cutting the coastal road and railway. This effectively puts the 164th and the Bersaglieri into a pocket, trapped against the coast. The two divisions report that they are also coming under attack from the east by the Australian 9th Division. Unless decisive action is taken they will be cut off and perhaps overrun. If you wish to order them to hold their positions and launch a counterattack by the 90th Light Division then go to **Decision 314**. If you want to authorise them to conduct a fighting withdrawal towards Sidi Abd el Rahman, covered by the 90th, then go to **Decision 335**.

269

0800 hours 27 October 1942

The Littorio Division moves up to cover the Kidney Ridge area. It seems readily apparent to you that the enemy is looking to build up and make a determined drive towards Tel el Aqqaqir. From the actions of the enemy it is also obvious that he is attempting to widen his hold on this central area in the north. From this salient they could strike north towards the coast, west towards the Rahman Track or swing to the south then west to sever your supply lines to your troops stationed on the southern edge of the front. You need to decide whether you will continue to hold this area or launch a counterattack. If you wish to launch a counterattack using the Littorio Division then go to **Decision 339**. If you think that it is more prudent to hold your positions for now then go to **Decision 360**.

270

2400 hours 28 October 1942

Bravely the 21st PZ Division drives into action against the leading prongs of the enemy armour. They manage to break through using up huge amounts of your valuable and dwindling fuel supplies. The attack is a partial success; large elements of Ramcke's parachute division, along with the Italian Bologna and Brescia divisions, have been saved. What remains of the 21st is now falling back towards Fuka. If you wish to hold at Fuka then go to **Decision 316**. If you think that Fuka is already compromised and you need to fall back further then go to **Decision 358**.

271

2400 hours 28 October 1942

The 21st Panzer and the Ariete Division launch a determined counterattack. It catches the enemy unprepared and over-extended. For several fleeting hours you hope that you can turn the tide of the battle and, perhaps, regain some ground. But the casualties amongst the armoured divisions are mounting and also your troops in the south are taking the opportunity to put some miles between them and the chasing enemy units. Several battalions of troops have managed to break through thanks to the sacrifices of your armour. They are trying to make their way north and west. The 21st PZ Division also cuts a path through to Fuka, allowing more troops to escape. Now go to **Decision 316**.

272

You take the decision to hold. Only the superhuman efforts of your staff and divisional commanders manage to organise a new defensive line. Your troops are woefully short on equipment, fuel, food and water. You can only hope for a miracle as the enemy formations begin to close in on your positions. Suddenly the sky darkens and a torrential rainstorm begins. The rain continues for hours; this is exactly what you needed. The soft sand to the south, your unprotected flank, is now impassable. You are able to contact

High Command and appraise them of the situation. Reluctantly they allow you to begin a phased withdrawal towards El Agheila, where you can establish a better defensive position. The rains have taken the momentum of the enemy offensive from them and you are able to continue to hold for several more weeks. Eventually you are given permission to withdraw into Tunisia, leaving Libya forever in February 1943. You leave with huge regrets, however. Had you been supported then you would have been able to have conquered Egypt, seize the Suez Canal and who knows, you could have had the opportunity to link up with German forces sweeping from the north. You could have captured the Middle East and brought this war to an end.

273

0200 hours 30 October 1942

Disastrous news comes far earlier than you could have possibly feared. The enemy has moved into El Daba in force. There is also a rumour that Fuka has been captured. Your western flank has almost completely disintegrated and your troops are cut off from the west. You have only two options; you can try to fight your way out and go to **Decision 318**. Or you can surrender what remains of your command at Ghazal and go to **Decision 297**.

274

0800 hours 27 October 1942

The Ariete Division moves into position to support. The front is stabilised to some extent, but it is still under pressure. The danger is now in the north, with the enemy seemingly concentrating on the Kidney Ridge area and the coastal strip. You feel that you need to turn your attention to this sector of the battlefield. Now turn to **Decision 382**.

275

The situation in the south is extremely dangerous and your front has been compromised. You are left in no position other than to order a retreat. The only question is how fast this retreat should be. If it is too fast then your foot-slogging Italian infantry units will be cut off. if it is too slow then your armour will be engaged and probably destroyed. If you want to order a slow retreat, then go to **Decision 341**. If you want to order a more rapid retreat then go to **Decision 362**.

276

The Ariete Division takes up its positions to the west, covering the flank from the Jebel Kalakh to the Qattar Depression. No sooner has the division got into position than reports start filtering in that the enemy probes and attacks have begun to tail off in the south. There is an unconfirmed report that the enemy armoured division positioned in the south, believed to be the 7th Armoured, has been moved to the northern sector. This undoubtedly means that the next major attempt at a breakthrough must come in the north. With the southern front relatively stable, you take the decision to move the 21st PZ Division north, as a mobile reserve, figuring that it can do far more good protecting the Rahman Track than sitting in reserve in the south. At the same time it is closer to its fuel supplies. Now turn to **Decision 382**.

277

Your armoured units begin their approach to the Kidney Ridge feature. They immediately come under artillery fire, breaking up their formations, but they press on. Suddenly enemy armour appears, focusing on the flanks of your formations. They are attacking from the north and from the south. You are in direct contact with von Thoma, who is personally directing the assault. He asks you whether your armour should engage with the enemy tanks, or whether the counterattack should be called off. If you want to order von Thoma to tell his tanks to engage then go to **Decision 254**. If you are fearful that your armour will be overwhelmed then go to **Decision 386**.

278

2400 hours 28 October 1942

Your troops are now back in their defensive positions. The enemy has been plastering your defensive line, but so far have failed to try to break out of the Kidney Ridge salient. News is now coming in of a renewed enemy push; a major attempt is being made from Point 29, slightly to the north of the Kidney Ridge salient. You can see from the map what the enemy is intending. They aim to cut off the 90th Light Division and the Bersaglieri by bursting through and cutting the coast road. If you want to order the troops in the coastal area to hold then go to **Decision 365**. If you want to give them permission to fall back towards Sidi Abd el Rahman before they are cut off then go to **Decision 344**.

279

2400 hours 28 October 1942

The enemy has preceded an assault from Point 29 with a tremendous artillery barrage. He is pushing armour and infantry west and northwest, with the aim to cut off your troops holding the coastal sector. Elements of the 164th Division and the Bersaglieri are being forced to give ground to prevent their positions from being overwhelmed. In the immediate term you only have limited resources to support them. If you want to order the 90th Light Division to move up and support the sector then go to **Decision 255**. If you think that the 90th Light has already taken enough punishment and they need time to reform then your existing troops will have to hold for now and you should go to **Decision 280**.

280

2400 hours 29 October 1942

The Australians continue to probe forward from the Point 29 position. However the assault in this area is far less intense. Your limited intelligence reports suggest that Montgomery has begun withdrawing units into a general reserve. This means that he aims to launch an assault somewhere along your front; probably in the northern sector. If you think his focus will be on Point 29 then go to **Decision 303**. If you think he intends to reinforce the Kidney Ridge salient and use that as a springboard for a new assault then go to **Decision 281**.

281

You decide that the pivotal sector in the north will be the Kidney Ridge feature. You order your armour to move up to cover this part of the front. There are limited enemy attacks along the coastal strip, but everywhere else is ominously quiet. Now turn to **Decision 368**.

282

2400 hours 31 October 1942

It has been a disastrous day for the 21st Panzer Division. They have been engaged in fighting for nearly twenty-four hours and have now been swept aside. The pocket has been sealed and the enemy is breaking through all along the coastal strip. The future of your entire army is now in jeopardy. There seems to be no cohesive line of defence along the coast. Enemy units are punching through, leaving behind them isolated pockets of resistance, which are being mopped up by their infantry. You need to make a decision about the coastal strip and whether to continue to try to resist against the odds. If you want to order renewed resistance to try to stem the tide then go to **Decision 367**. If you think it is time for you to order your troops in the north to fall back along the coast and hope that as many of your troops as possible in the south can extricate themselves then go to **Decision 389**.

283

You reluctantly order the withdrawal back towards Fuka. Each mile that you retreat means leaving more men and equipment behind to fall into the hands of the enemy. It is a terrible decision to make; abandoning men that have fought for you for so many months. The situation at Fuka is desperate; many of the units are completely out of fuel. Some of the battalions are reduced to less than company strength, and whole divisions appear to have evaporated, having been cut off or captured. You know that you need to withdraw further west, but this will need Hitler's personal permission. You know that you will have to wait for him to deliberate and there may not be the time to allow that process to take place. If you want to order the army, or what remains of it, to retreat further west with immediate effect then go to **Decision 305**. If you want to wait for Hitler's permission to fall back then go to **Decision 326**.

284

1000 hours 2 November 1942

Under cover of a rolling barrage, enemy armour supported by strong infantry units have punched their way through to the Rahman Track. Your line facing the Kidney Ridge salient has virtually disintegrated and there is chaos everywhere, with reports of enemy units roaming at will to the rear of retreating German and Italian units. Some divisions report enormous losses and others state that whole battalions have been cut off. This is obviously the main effort launched by Montgomery. He clearly aims to break through all the way to the Telegraph Track and then swing north to cut off your troops in the south, whilst surrounding those in the coastal strip. You have only two options. You can commit your armour to the fight and try to blunt the enemy assault by going to **Decision 286**. Alternatively you could save your armour as your only hope of trying to re-establish a cohesive front further west. As the minutes pass you realise that this new line must probably be no further east than El Daba. If you want to choose this option, then go to **Decision 307**.

285

After considerable deliberation you decide that you will order a withdrawal. At least this will preserve what remains of your armoured force and, hopefully, the bulk of your troops in the north will be able to execute a disengagement and general shift to the west. What worries you is the situation in the south, as a withdrawal will undoubtedly mean that the bulk of those troops will be cut off and lost. The situation is deteriorating minute by minute and you need to make a decision as to where to draw your new defensive line.

If you think that it is feasible to establish a new position at the end of the Telegraph Track, at El Daba, then go to **Decision 262**. At least this choice would give your troops in the south the maximum opportunity of retreating northwest to link up with you. Alternatively, the other defensive position could be Fuka, which is further west. If you wish to choose this option, then turn to **Decision 287**.

286

Your armour rolls into battle, perhaps for the last time, in a desperate attempt to stem the enemy tide. General von Thoma keeps you abreast of the situation, reporting successes and failures as the last of your armour engages the lead units of two enemy armoured divisions. You press him for his opinion of how the battle is developing. Uncharacteristically he is unsure and merely says that it still hangs in the balance. There is a pause as the enemy begins to regroup. You know that this next attack may well overwhelm what remains of your armour. Von Thoma agrees with you. If you want to pull the remnants of your armour out of the line before the next attack then go to **Decision 329**. If you want to keep your armour in position then go to **Decision 308**.

287

2400 hours 3 November 1942

The bulk of your mobile troops have been able to reach Fuka. The men have arrived on every conceivable mode of transport. Your army bears the resemblance of a travelling circus; a mixture of buses, trucks, cars and even mule carts. There is a running fight around the Rahman Track and the Telegraph Track, as lead enemy units encounter large retreating groups of Italian infantry. Many of the Italian units appear to be determined to break through to the northwest and reports suggest that the fighting is fierce. Intelligence also suggests that the enemy 7th Armoured Division is moving at speed along the coastal road. Further south the enemy has pulled its two other armoured divisions away from the conflict with the Italian troops and has swung them northwest, with the aim of linking up with the 7th Armoured. You now need to make a decision as to whether you can make a stand here at Fuka. If you think that you should attempt to make a stand then go to **Decision 331**. If you feel that you need to fall back further, to Baggush, then go to **Decision 352**.

288

The enemy floods into El Daba and presses on west, overwhelming Fuka. Your troops continue to fight in an ever-more desperate attempt to evade the oncoming enemy. The only hope that you have is that Montgomery over-extends his supply lines. What worries you the most is the fact that you have no protection on your southern flank. All semblance of resistance has disintegrated and at any time the enemy could race columns across the desert to the south and then swing north to cut you off. If you feel that the fight is now over and that you should halt your men, end the slaughter and begin surrender negotiations then go to **Decision 397**. If you think that you can still put up a fight and make the enemy pay then go to **Decision 318**.

289

0800 hours 3 November 1942

The first piece of paper that you are handed is a polite note from the commander of the Ariete Division. He tells you that with regret his troops have ignored your order refusing them permission to help the Italians in the retreat. Instead, he has driven what remains of his force southeast and has negotiated the surrender of the Italian contingent to the enemy. You have made a terrible mistake and you know that there will be hell to pay. If you wish to tender your resignation as commander of the Afrika Korps then go to **Decision 309**. If you wish to continue to fight on regardless then go to **Decision 351**.

290

You decide that you have to fall back to Mersa Matruh before it is too late. You arrive ahead of the bulk of what remains of the Afrika Korps and send a report, giving your current situation. You confidently expect that Hitler will demand your resignation. At the very least he will replace you. Incredibly he chooses to keep you as commander of the Afrika Korps. At least now you can try to salvage as much of your force as possible and continue to struggle against the insuperable odds in North Africa. Luckily, heavy rains save you from the worst possible outcome; encirclement. The torrential downpour has turned the soft sand in the south into an impassable, impenetrable vastness that the enemy cannot exploit. You can now carefully begin your withdrawal, reducing your supply lines and continue the fight in North Africa.

291

You touch down in Athens and make a short journey to the German headquarters on Syntagma Square. There is a relatively small occupation force in Greece and you know that they have insufficient troops to provide you with any reinforcements. You investigate the possibilities of flying in supplies via Athens and Crete. The Luftwaffe explains the difficulties and the fact that surface vessels and aircraft are routinely intercepted by RAF aircraft operating out of Egypt and Cyprus. In any case, large numbers of their transports have already been sent north and are vital to the supply of German troops on the Russian front. They also complain that the infrastructure, particularly the airfields, are inadequate, but you remain unconvinced. If you want to fly direct to Libya then go to **Decision 354**. If you want to fly via Crete to refuel and to assess the ground situation and logistics there yourself then go to **Decision 333**.

292

1500 hours 26 October 1942

With an incomplete intelligence report you cannot predict exactly what your counterattacking force is likely to encounter should they try to retake Point 29. You can see why Montgomery has moved to take it. It is a pivot point from which he can launch a determined assault towards the coast, in the hope of taking Sidi Abd el Rahman, at the end of the Rahman Track, which is your main supply route. Montgomery is still an unknown factor; is this a feint, or is this his main effort? If you think that the north is likely to be the main area of operations then go to **Decision 334**. If you think there is a chance that he could be trying to fool you into weakening your southern front then go to **Decision 313**.

293

2400 hours 26 October 1942

You authorise the two divisions to pull back towards Sidi Abd el Rahman. They form a new defensive line, but it is unlikely that they will hold for any length of time without support. The coastal road is crucial; if this falls then your ability to resupply, or in the worst case withdraw, your army to the west will be at best compromised and at worst impossible. Your mobile reserves to the west consist solely of the Trieste Motorised Division. Intelligence sources suggest that the British 1st Armoured Division has already begun making probing attacks against the new defensive positions around Sidi Abd el Rahman. Your choices are stark; you must either tell the two divisions to hold and give you time, in which case go to **Decision 356**. Alternatively you can reinforce them now with the 90th Light Division and go to **Decision 377**.

294

1100 hours 27 October 1942

For the past couple of hours the 90th Light Division has launched a series of counterattacks against the enemy front along the coastal strip. Unfortunately the enemy has been able to slip significant amounts of armour through the minefields, backed by infantry and covered by artillery fire and determined aircraft cover. The pocket containing the 164th and the Italians has also been attacked from the east by the Australians moving along the coast. If you want to order the 90th to launch another counterattack then go to **Decision 336**. If you want to order them to fall back then go to **Decision 315**.

295

2400 hours 29 October 1942

Arriving in Mersa Matruh you re-establish contact with Berlin. You try to explain the situation to the German High Command. It appears that Hitler is already aware of the disaster that has befallen the Afrika Korps. You are not giving an opportunity to explain the situation. Hitler's orders are that you are to be replaced with immediate effect. You are relieved of command of the Afrika Korps. Your orders are to return to Berlin and with a heavy heart you bid farewell to your staff. You head for the nearest airfield and to a very uncertain future.

296

2400 hours 27 October 1942

You order your troops in the south to begin a series of phased withdrawals. The 21st PZ Division, along with the Ariete Division, still remain strong enough to hold the main routes open. Your divisions in the south are falling back along the Telegraph Track. You are concerned that the enemy will try to flank you to the south. The Telegraph Track meets the coastal road near El Daba, but this may be too far east for you to establish a new line of defence. Already the enemy is pushing along the coastal route and this may have to be kept open. If you want to order a retreat towards Fuka, with the 21st Panzer and the Ariete holding the route open, then go to **Decision 271**. If you want to order your troops to hold further east then go to **Decision 338**.

297

You are driven towards the front line, where you order an aide to go forward on a motorcycle to open negotiations for a ceasefire. Impatiently you wait for nearly an hour before he returns. The ceasefire has been agreed, although the enemy's instructions are that you are to hold in place and make no attempt to retreat towards the west. It is all that you can do to save what remains of your tattered command. There is little point in shedding more blood in a futile attempt to continue the struggle. The following day you meet with Montgomery himself in no man's land and sign the surrender papers. You fear for the fate of your family and hope that Hitler's wrath will not take them.

298

The Free French are working their way through the gaps in the line. They are clearly trying to establish a path through the minefield. Reports seem to suggest that enemy armour is organising itself in order to attempt a breakthrough. You now realise that the Pavia Division may be in grave peril. If you want to order them to continue to hold then go to **Decision 340**. If you want to reconsider, allowing them to fall back, then go to **Decision 319**.

299

1400 hours 28 October 1942

You order a fighting withdrawal. The enemy is being uncharacteristically aggressive and is pushing armour and infantry forward. They can sense a degree of panic, particularly in the isolated Italian units. The Italians in their forward positions will find it incredibly difficult to withdraw in any kind of order, as they lack transport and logistics. This could easily turn into a disaster. You need to make the decision as to whether you will attempt to slow down the withdrawal, possibly sacrificing your armour, in order to save the Italians, or whether to pull your mobile reserves out before they are destroyed. If you want to order a slow and deliberate retreat, giving the Italians the maximum opportunity to extricate themselves then go to **Decision 341**. If you want to order a more rapid withdrawal then go to **Decision 362**.

300

Your armour closes in on the Woodcock position, with the 21st Panzer Division leading, supported by Italian armour. Elements of the 15th Panzer are also able to offer their support. Your assault directly coincides with a major counterattack by British armour. Von Thoma, who is directing the assault, reports that upwards of 150 British tanks are moving to engage. If you want to order von Thoma to send his tanks in to engage the enemy armour then go to **Decision 343**. If you want to pull your tanks back and try to hold off the enemy assault with your anti-tank guns then go to **Decision 364**.

301

You decide to press on with your counterattack. The 21st PZ Division falls back into reserve to refuel and re-arm, with the cutting edge now taken over by the 15th PZ Division. The British are falling back in the face of this determined assault, but you are fearful that you might overstretch yourself. If you wish to continue counterattacking, then go to **Decision 323**. If you wish to call off the attack and fall back to your original defensive positions then go to **Decision 278**.

302

1800 hours 27 October 1942

The 15th PZ Division rumbles into action. The shock of the reserves shatters the British advance. Von Thoma reports that enemy armour is falling back, with huge losses. The salient has been turned into chaos, with scattered British units holding out in places and retreating in terror in others. This may be your chance to deliver a fatal blow to Montgomery. If you want to order the counterattack to continue, then go to **Decision 323**. If you think that you have taught the enemy a lesson and that you should preserve your armour, then go to **Decision 278**.

303

Having examined the map and tried to predict the enemy intentions, you are certain that Point 29 will be the launch pad for any renewed enemy assault. You move up your armoured reserves to cover Point 29. Although your armour is considerably weaker than that of the enemy, you hope that it will be sufficient to counter any breakthroughs. It seems clear to you that Montgomery intends to break through and seize the towns along the coastal strip. This will deny you the opportunity to resupply your troops and to fall back in any kind of order should you need to consider a general withdrawal from the battlefield. Now turn to **Decision 368**.

304

2400 hours 31 October 1942

There are now confirmed reports that the enemy 1st and 10th Armoured Divisions are advancing in force along the coastal strip. Every attempt that your troops have made to establish a new line has simply led to their being overrun. The whole of the coastal area is now in grave danger and with it your hope to be able to pull any of your troops back towards the west. Now very little stands in the way of a total victory for the enemy. You still have one last throw of the dice. If you want to commit the last of your armoured reserves to try to stop the tide of enemy advances then go to **Decision 325**. If you think that this would be a useless waste of your most valuable resource and that the infantry and other units must fend for themselves then go to **Decision 346**.

305

You decide to order the retreat now, before it is too late. You re-establish your command centre and almost immediately receive a coded message from Hitler himself. He is absolutely livid with you and blames you for having lost the army in North Africa. He tells you that on every other front the German forces are victorious and advancing. You are to hand over your command to General von Thoma with immediate effect and you are to present yourself to Hitler in Berlin. Undoubtedly this is the end of your military career, but your fate may actually be worse.

306

You take the decision to move your armour south before it is utterly destroyed by the enemy bombardment. As soon as he receives your orders von Thoma tells you that the fuel situation is critical. He has barely enough fuel for the armour now and virtually none in reserve. This could be the only move you can afford for the armour. It is a difficult decision to make; leave them in position and allow them to be destroyed by enemy artillery, or move them south and risk them running out of fuel. If you want to leave them in position then go to **Decision 284**. If you want to move them, then go to **Decision 369**.

307

Your armour moves northwest towards El Daba, in the hope that it can establish a firm defensive line. You know that this force will be inadequate and that it cannot possibly hold out indefinitely. The question is whether you should remain here at El Daba, or consider falling back. Of more immediate concern is whether to order your whole army to withdraw to the northwest or whether to order the bulk of them to remain in position in the hope that you can organise a counter offensive. If you want to order your army to fall back towards the northwest then go to **Decision 393**. If you want them to remain in position then go to **Decision 263**.

308

1800 hours 2 November 1942

By dusk the battle is over. Having committed your entire force to the engagement you have been able to beat off the enemy attacks. It is, however, a Pyrrhic victory, but a great moral boost. The enemy's westward sweep has been curtailed for now. Your troops are able to move into defensive positions. If you wish to re-establish the line, knowing that it may be a temporary move, then go to **Decision 349**. If you feel that you need to use this breathing space as an opportunity to withdraw then go to **Decision 372**.

309

You sit down at your map desk, having dismissed your staff. You need a moment to think as you compose your resignation. You call in the radio operator and dictate your message to Berlin. The reply is uncharacteristically prompt. Army command instructs you to make for the nearest airfield and to leave the theatre, heading west. They will organise an aircraft to fly you across the Mediterranean. You are to report to Berlin. The days of the Desert Fox are over and all you can hope is that you are allowed to slip into obscurity and are not forced to pay the ultimate price for your failure.

310

0800 hours 3 November 1942

You wake up to the news that the Ariete Division has pushed several miles southeast, along the Telegraph Track, and has been able to extricate several hundred Italian infantry. They are now beginning to fall back, as the lead enemy units begin to close in on their position. They are also threatened from the north, where enemy armoured columns are roaming across the desert. Elsewhere, however, the enemy pursuit is slow. Perhaps Montgomery is outstripping his supply lines? This will give you a short opportunity to consider your strategy. Now go to **Decision 351**.

311

Your troops, made up of scratch units from dozens of different formations, launch a counterattack against the enemy columns racing west. You are able to hold open the coastal road behind you for now, but you need to make a quick decision as to whether you need to fall back towards Baggush and go to **Decision 352**, or wait, hoping that your stinging counterattack has deterred the enemy columns and go to **Decision 353**.

312

The aircraft touches down at a blustery Belgrade. This part of occupied Europe is becoming increasingly dangerous, with numerous partisan raids holding down huge numbers of German troops and their allies. This seems to be another theatre that is receiving priority over your operations in North Africa. Your aircraft is refuelled and ready to leave, but engine problems delay you. You are desperate to get back to the front, wondering how Stumme is coping with the situation. The reports that you have received so far, even the updates that have been routed to Belgrade, are patchy and contradictory. Eventually your aircraft is ready and you make the perilous crossing of the Mediterranean, heading for Libya. Now go to **Decision 354**.

313

You decide to leave the 21st PZ and the Italian Ariete armoured division in the south for now, but take the precaution of ordering the 90th Light Division forward as a covering force in the north. Your intelligence officers report that the enemy has launched the 44th Infantry Division and the 7th Armoured, along with Greek and Free French troops, against the Italian 185th Folgore Division in the south. Perhaps you were right and this is where Montgomery hopes to break through? The Italian division has taken enormous casualties and it does not appear that they will be able to hold for much longer. If you want to support the Folgore Division with the Ariete armoured division then go to **Decision 274**. If you want to give them permission to fall back to the next line of defence then go to **Decision 249**.

314

0800 hours 27 October 1942

There has been vicious fighting throughout the night. You are now beginning to regret not assembling all of your motorised units in the north to throw the British back towards their start positions, with concentrated and properly-planned counterattacks. You have taken unacceptable losses from air and artillery bombardment. Nonetheless the 90th Light Division has managed to make some headway, but now, during daylight, they are coming under renewed air assault and pinpoint artillery bombardment. Now turn to **Decision 294**.

315

1200 hours 27 October 1942

Your order to fall back has compromised the Rahman Track. Scattered elements of the 164th and the Bersaglieri have managed to fall back, having broken through with the help of the 90th Light Division. Now the situation is becoming very dangerous, as the enemy is flooding forward in the north and there is no line to prevent them from doing so. You need to quickly make a decision. If you want to order the remaining troops in the coastal area to turn, hold and fight then go to **Decision 378**. If you believe that the situation is now lost and you need to order a withdrawal of your army then go to **Decision 357**.

316

2400 hours 29 October 1942

You fear that the enemy will simply flank you to the south. Your position is extremely precarious. However retreat is not a decision that you feel able to make without first seeking permission of High Command, or perhaps from Hitler himself. You send a coded message requesting permission to withdraw further to the west. You have no idea how long it will take for a reply to come back from him. As you read the reports and stare at the situation map you realise that you may not have the time to wait for his reply. If you do want to wait for permission to retreat towards the west then go to **Decision 337**. If you think that for the sake of your men you need to order a retreat west now and take full responsibility for it then go to **Decision 358**.

317

2400 hours 27 October 1942

It does appear that the southern part of your front is relatively stable. The enemy has been launching probing attacks but so far your units have been able to hold their positions. There may be an opportunity for you to shift your two mobile units in the south to the north. If you would like to move the 21st Panzer Division and the Ariete Division to the northern sector then go to **Decision 359**. If you think this is not a good decision and that they should remain in position in the south then go to **Decision 381**.

318

Your troops launch a desperate counterattack towards the west. Your army, which once moved with such precision across the desert, has been reduced to a rag-tag collection of disparate companies mixed together. They are still a determined and effective fighting force and it is still with some pride that you see them battling their way west. Now turn to **Decision 247**.

319

The Pavia Division is attacked throughout the day and during the course of their withdrawal. Their line is very thin and poorly protected. You may need to send additional troops to the south to bolster their position. You spend several hours considering the situation and then, against all odds, new intelligence arrives. It tells you that the attacks in the south have stopped. The enemy has downgraded to localised actions only. It is also unconfirmed that the enemy's 7th Armoured Division, which was stationed in the south, has been shifted to the north. You immediately take the decision to follow suit and move the 21st Panzer Division to the north. Now turn to **Decision 382**.

320

2400 hours 28 October 1942

The British 7th Armoured Division has burst through the Jebel Kalakh sector of the southern front. There is worse news in the north; two further enemy armoured divisions, supported by significant numbers of infantry, have reached the Rahman Track from the Kidney Ridge salient. Local commanders are trying to extricate their troops before they are trapped in pockets behind enemy lines. Elements of several of your divisions are falling back on Ghazal, along the coast in the north. Now turn to **Decision 248**.

321

1600 hours 27 October 1942

Intelligence reports suggest that the Snipe position, in the southwest of the salient, is occupied by the enemy Rifle Brigade and a number of anti-tank guns. They will not be easy to overwhelm. The attack is led by the 21st Panzer Division, supported by the Littorio armoured division. Enemy artillery opens up soon after your tanks leave their start positions. As the distance between your lead elements and the objective shorten your troops come under anti-tank gun fire and begin to take casualties. As the distance shortens more and more of your precious armour is lost. Von Thoma contacts you. He is directing the assault himself. Enemy armour is moving up to support the Snipe position. He needs you to decide whether to call the attack off or press on. If you want to call the attack off then go to **Decision 385**. If you want to continue the attack, then go to **Decision 253**.

322

1400 hours 27 October 1942

The 90th Light Division moves up to attack. Heavy enemy artillery and aircraft attacks break up their deployment before they can even get into action. It is obvious to you that the enemy has anticipated your move and that any attempt to attack Point 29 would be suicidal. This means that you need to refocus and perhaps consider launching an assault elsewhere along the front. If you want to attack the Kidney Ridge feature with your armour then go to **Decision 277**. If you want to attack the northwest of the enemy salient then go to **Decision 300**. If you want to attack the southwest of the salient then go to **Decision 321**. Alternatively you can remain on the defensive and await development, in which case you should go to **Decision 279**.

323

The enemy has somehow regained the resolve to resist. The advances by the 15th PZ Division are becoming less noticeable and your casualties are mounting. The 21st PZ Division has now been refuelled and re-armed. It is now time to make a crucial choice. If you want to continue to push forward in the hope that you can break the enemy and regain the initiative in North Africa then go to **Decision 387**. If you think that it is too risky and that your armour will inevitably be overwhelmed and you need to return to the defensive then go to **Decision 256**.

324

2400 hours 20 October 1942

Under a barrage of artillery fire the Australians launch another attempt to break out towards the coast. In grim and desperate fighting the Australians manage to push through to take the coastal road. This has made the position of the 164th Division and the Bersaglieri extremely precarious. Effectively they are in a salient, almost a pocket, cut off from the bulk of your army. Of particular worry is a position that the enemy has codenamed Thompson's Post. You can now see that in holding this position the enemy is able to deny you the opportunity to pull your troops out of the pocket, or to reinforce them. Thompson's Post needs to be taken back. Effectively you have two choices. You can launch the 90th Light Division at the enemy position and turn to **Decision 345**, or alternatively you can move the 21st Panzer Division into the sector and send them against the enemy position at Thompson's Post by going to **Decision 366**.

325

You commit your armour to a major tank battle in the coastal area. Your experienced tank crews inflict heavy damage on the lead enemy units, cutting off their probing attacks and overwhelming exposed positions. For a time the enemy is held; your counterattack has caught them unawares. But then you receive word that the enemy has committed a third armoured division. Now you fear for your armoured units. As the minutes pass the situation is beginning to deteriorate as your own losses mount. The enemy is renewing its forward movement, pushing through your thinning ranks and positions. You know that this is perhaps the key moment in the whole battle and that whatever you decide now will determine the fate of your army. The question is whether your armour is sufficiently strong to withstand the onslaught of three enemy armoured divisions. Their strength is considerable, outnumbering you at least four or five to one. You must decide whether to continue to resist and trust in the experience of your tank crews. If so go to **Decision 367**. Alternatively you must order your armour to withdraw to the west before it is overwhelmed and go to **Decision 389**.

326

Already you are regretting the fact that you have waited this long for a reply from High Command and from Hitler. The enemy has already flanked you and your intelligence suggests that he is making for Baggush, on the coast. A coded message finally arrives, signed by Hitler. He agrees that you can make a limited retreat to the west, although he orders you to stabilise the front as your sole priority. However it may be too late to make any valuable decision. If you want to resign your command then go to **Decision 379**. If you want to order an attempt to break through to Mersa Matruh then go to **Decision 247**.

327

1100 hours 2 November 1942

Your armour begins to move into position to engage the enemy tanks. It takes tremendous fire from enemy artillery and aircraft as it manoeuvres into position. Von Thoma reports that elements of at least two enemy armoured divisions are pushing forward. Casualties are mounting as von Thoma keeps you updated with developments. You can sense that the counterattack has failed. After just a few minutes of close engagement you have lost around 100 tanks from artillery fire and anti-tank fire. Your gamble to use the armour has failed; the sheer weight of fire power commanded by the enemy in this sector has overwhelmed them. Now turn to **Decision 391**.

328

With great reluctance you decide to call off the counterattacks around Tel el Aqqaqir. You feel that nothing can be gained in sacrificing your armour in a battle that you cannot possibly win. You order von Thoma to pull the armour out of range, moving it west into new positions. Von Thoma tells you that the fuel situation is extremely critical. His armour has not received fuel supplies for nearly thirty-six hours and his fuel bowsers are empty. He doubts whether there is sufficient fuel for the tanks to engage in another battle. If you want to order von Thoma to hold the line and give your troops in the south sufficient time to fall back towards more secure positions then go to **Decision 349**. If you want to order von Thoma to pull the armour back to the northwest, using up most of the remaining fuel, then go to **Decision 285**.

329

What remains of your armoured force retreats from the battlefield. They are seriously depleted in numbers. The tank crews are halting to siphon off fuel from abandoned vehicles. Back on the battlefield your infantry and anti-tank weapons have been largely overwhelmed. The armour, now in reserve, is not yet safe. Was your armour originally positioned on the coastal strip? If so go to **Decision 350**. If it was originally covering the Kidney Ridge salient you should go to **Decision 371**.

330

You decide that you should continue to fight on. You begin falling back towards Mersa Matruh. As the hours pass the number of your troops making it back dwindles. This gives you the certain knowledge that the enemy is close. Precious few of the Afrika Korps have survived this disastrous campaign. Nonetheless you are amazed to see them digging in around Mersa Matruh, determined to hold back the enemy at all costs. You know that the enemy can work their way around your southern flank but then a miracle; torrential rain. You know that the soft sand to the south has been rendered impassable. The enemy, with their superiority in transport and armour, will now be restricted to trying to burst through on a very narrow front. This gives you the opportunity to regroup. You begin by sending your least mobile troops west, towards El Agheila. Finally you give permission for your mobile troops to follow; they are shadowed all the way by Montgomery's armour. All of the impetus of his assault has been dissipated and it is not until early February 1943 that you are finally forced to withdraw from Libya forever. As the last elements of the Afrika Korps slip across the border into Tunisia you know that the end in North Africa is close.

331

You hold Fuka for as long as possible, even sending probing forces to the southeast to help link up with bands of exhausted Italian infantry, who have hiked miles across the desert to link up with you. The decision to do this has been worthwhile. You talk the situation over with your senior commanders. The general consensus of opinion is that you should fall back towards Mersa

Matruh. You know the terrain extremely well and you know that soft sand to the south will protect your flank. The only other option is to continue to try to hold further to the east. But your new front is perilously exposed to a flanking move by the enemy. Your commanders have mixed opinions and ultimately the decision has to be yours. If you want to remain in position then go to **Decision 394**. If you think that you need to pull back toward Mersa Matruh then go to **Decision 264**.

332

Hitler is absolutely livid with your decisions and he also feels that he has been personally embarrassed, as the Italians blame both you and him for their enormous losses. Hitler demands your immediate resignation. As you read the message you realise that it is not as straightforward as this. You owe your men your leadership to get them out of this terrible position and get them safely to Mersa Matruh. If you wish to do this then go to **Decision 380**. If you feel that resignation is your only option then go to **Decision 379**.

333

You touch down at Heraklion. It is clear as you step out of the aircraft that the scars of the battle that took place in May last year on Crete have not yet healed. Vast numbers of German paratroopers died to take this island yet now it seems a sleepy backwater and the plans to use it as a springboard to attack Egypt and Cyprus have been long forgotten.

The aircraft is quickly refuelled but the Luftwaffe cannot provide you with any escort. Your Ju52 takes off, crossing the mountains and heading into the Libyan Sea. Suddenly an RAF fighter bomber emerges from a low cloud. It strafes the fuselage of your transport and arcs away to make a second pass. The pilot, unsure as to continue or turn back, drops down to sea level, hoping to elude the raider. Your aircraft is spotted and the next attack riddles the cockpit and the Ju52 tumbles into the sea. You were never destined to return to Libya to take command it seems.

334

2000 hours 26 October 1942

You decide to move up the 21st PZ Division to the north and bring the 90th Light Division forward. Meanwhile the Italian Littorio Armoured unit shifts to the southeast to help cover the Italian Bologna Division, which in turn is covering the Ruweisat Ridge, another pivotal point in your defensive line. Throughout all these moves your formations have come under persistent air attacks. You are beginning to realise that the only safe time to move may be during the hours of darkness. Now turn to **Decision 382**.

335

0800 hours 27 October 1942

Reluctantly you give the two divisions the permission to conduct a fighting withdrawal towards Sidi Abd el Rahman. The enemy has harried and probed along the coastal road all night and has closed with the new defensive perimeter that is being hastily set up. In a frank discussion with the commander of the 164th he admits that it is unlikely that they will be able to hold without serious reinforcement. Unfortunately your options are relatively limited. You can either throw in your last major reserve unit, the 90th Light, to stabilise the front and go to **Decision 377**, or you can order the commander to hold at all costs to give you time to consider your options by turning to **Decision 356**.

336

1200 hours 27 October 1942

You have managed to stabilise the front along the coastal strip to some extent, but the situation is still extremely dangerous and it gives you grave cause for concern. You know that at any time the severe enemy pressure along the coast may well cause the whole front to collapse. If this happens the enemy will race to capture the coastal towns, cutting off your army from any hope of orderly retreat. If you want to order your army to begin a withdrawal to the west before it is too late then go to **Decision 296**. If you do not wish to do this and consider other options then go to **Decision 317**.

337

2400 hours 30 October 1942

Already you are regretting the fact that you have waited this long for a reply from High Command and from Hitler. The enemy has already flanked you and your intelligence suggests that he is making for Baggush, on the coast. A coded message finally arrives, signed by Hitler. He agrees that you can make a limited retreat to the west, although he orders you to stabilise the front as your sole priority. However it may be too late to make any valuable decision. If you want to resign your command then go to **Decision 379**. If you want to order an attempt to break through to Mersa Matruh then go to **Decision 247**.

338

2400 hours 28 October 1942

The enemy has pushed what you believe to be the 7th Armoured Division through a gap in your line. They have crossed the Rahman Track and it appears that they have probed towards the Telegraph Track. This is a disastrous development for you. Your only options are to turn and fight or to order your troops to fall back towards Baggush, on the coast. If you want them to turn and fight then go to **Decision 248**. If you want to order a withdrawal towards Baggush then go to **Decision 316**.

339

The Littorio Division fights for its life as it is assailed by two enemy armoured divisions. The Italian tanks are no match for the enemy armour yet the Italians continue to resist. But it is an unequal struggle. Brushing the Littorio aside the enemy pushes through towards the Rahman Track. Your worst nightmare has been realised; Sidi Abd el Rahman has been flanked. As well as that, your main supply and retreat route from the south has been cut off. You are left with only one option, which may well mean the sacrifice of the 21st Panzer and the Ariete Division. If you want to order these two mobile units to launch a counterattack then go to **Decision 271**. If you feel you should preserve them and try to establish a holding position further west then go to **Decision 246**.

340

0800 hours 27 October 1942

The Pavia Division has been flanked. The Italian infantry, lacking transport and panicked by enemy troops around their flanks and rear, is beginning to fall back towards Jebel Kalakh in the north. This has exposed the Folgore Division's flank. The enemy is advancing in some force with a hook heading west then north. You need to try to re-establish the line with the Jebel Kalakh feature as the pivot, or consider counterattacking with your mobile troops. If you want to try to re-establish the line, then go to **Decision 361**. If you want to launch a counterattack with the 21st Panzer and the Ariete Divisions then go to **Decision 383**.

341

As best you can, you keep abreast of the slow fighting withdrawal in the south. Even with your mobile forces attempting to hold to allow the foot slogging Italian troops to pull back to the northwest, there are still enormous losses. Your troops have launched localised counterattacks in attempts to break out of the trap, but you know that there is little that can really be done to save the bulk of the infantry. Nonetheless, your men are performing miracles against considerable odds and what can be salvaged is in full retreat to the northwest. Now turn to **Decision 248**.

342

The enemy begins a heavy bombardment of your positions around the Kidney Ridge area. You expect it to be followed by a major assault, aiming to cut the Rahman Track further west. What perplexes you is that the enemy is simply content to establish themselves in the salient but do not appear to be building up for a break out. This can give you an opportunity to seize the initiative. If you want to continue to hold the positions then go to **Decision 279**. If you want to launch an assault against Point 29, which has been taken by the Australians further to the north, then go to **Decision 322**.

343

Your armoured units are engaged in a major tank battle. The enemy is feeding more and more tanks into the conflict. Von Thoma is very concerned; although your armour is inflicting heavy casualties on the enemy your losses are also mounting. Even more alarming is the fact that your tanks are running out of fuel and their ammunition is in short supply. You could throw in the 15th Panzer Division as your lead element, as they are fresher. If you wish to do this, then go to **Decision 302**. If you wish to pull back and go on the defensive then go to **Decision 278**. If you want to pull back and launch an attack by the 90th Light Division against the Australians holding Point 29 to the north then go to **Decision 322**.

344

The 164th, the 90th Light and the 7th Bersaglieri begin falling back towards Sidi Abd el Rahman. They are able to stem the enemy tide and begin to re-establish a front. The enemy attack seems to halt temporarily, with the focus on the railway, which runs along the coastal strip. The enemy advance slows. If you want to leave the 90th Light Division in position in case of a renewed assault then go to **Decision 324**. If you want to move the 90th back into reserve then go to **Decision 280**.

345

Under cover of darkness the 90th Light Division moves forward. It seems that the enemy has anticipated the fact that you would launch a counterattack and has set up a rolling barrage, which straddles your forward positions and assembly areas. You are not surprised when you receive news that the 90th's attempt to counterattack and retake Thompson's Post has failed. The salient itself is under enormous pressure both from the south and from the east. Your troops are being forced to give ground. You need to make a decision.

If you want to order the troops in the salient to evacuate and re-establish a defensive perimeter around Sidi Abd el Rahman then go to **Decision 257**. If you want to order the 90th to try to coordinate a renewed counterattack with the troops in the salient then go to **Decision 388**.

346

You decide not to commit your armour to the battle but instead you order it to begin a withdrawal to the west, in an attempt to cover your slower moving units trying to extricate themselves from an ever-more dangerous position. The enemy takes advantage of the situation; deep thrusts of armour, supported by infantry, are roaring along the coastal road. At least two different enemy armoured divisions have been reported to be active around Sidi Abd el Rahman. A determined resistance of the town ends in failure as remnants of at least two of your divisions begin to stream west, just ahead of the lead units of the enemy armour. The next logical point of resistance is Ghazal. You may be able to hold the line there, perhaps long enough for more of your troops to escape. If you want to order your troops to try to form a new defensive line at Ghazal then go to **Decision 367**. If you think that Ghazal is already compromised and that you need to fall back even further along the coast, perhaps as far as Fuka, then go to **Decision 389**.

347

Whilst you desperately try to organise a new defensive perimeter around Fuka the enemy has established himself at Baggush and cut off the remnants of your army in a pocket on the coastal strip. You receive word that you are to evacuate Fuka and leave the army in the command of von Thoma. Hitler is now convinced that the battle has been lost and will not contemplate the prospect of a German Field Marshal falling into enemy hands. The rest of the army can be sacrificed as far as Hitler is concerned. They have failed him and so have you. An aircraft is to be flown in from Crete as soon as possible to deposit you in Berlin. You cannot expect a hero's welcome on your arrival.

348

1100 hours 2 November 1942

Your anti-tank guns scream, shattering the British armoured thrusts. They hammer the enemy armour. But over the next hour or so the combined force of artillery bombardment, aircraft attack and determined British tank attacks sees thirty-five or more of your key anti-tank gun positions overrun. The enemy armour has punched through the remnants of your defensive line and is streaming towards Tel el Aqqaqir. You know that this is a crucial point of the battle. All now hinges on your next decision. If you want to commit your armour to try to blunt the enemy attack then go to **Decision 261**. If you want to order the armour to withdraw and give a general retreat order for your troops then go to **Decision 285**. This may be your last opportunity before your front is shattered.

349

You impress upon each and every divisional commander the desperate situation that the army finds itself in. You order them to re-establish a line of defence and to prevent units from streaming northwest. Incredibly, your prompt action begins to stem the tide of retreats in the north, allowing the bulk of your troops in the south to fall back towards the Telegraph Track. Although you have lost a considerable amount of ground you begin to feel more confident that the army can be saved and that a new defensive line can be established. You have, however, lost the relative protection of the minefield belts and many of the positions are exposed in the desert. They will be incredibly difficult to resupply or reinforce. You realise that this temporary respite will be fleeting and that a more permanent decision about the positioning of your army will have to be taken. Now turn to **Decision 351**.

350

Having moved your armour from the coastal strip to the south, and now expecting it to be able to extricate itself by moving northwest, you have exhausted the bulk of the fuel. General von Thoma reluctantly reports to you that his tank crews are being forced to abandon their vehicles along the road and that there is nothing he can do to salvage them. There is absolutely nothing to protect the retreat of the rest of your army. You now feel that you have no option but to tender your resignation as commander of the Afrika Korps, having hopelessly mismanaged the battle. Now go to **Decision 379**.

351

2400 hours 3 November 1942

You talk the situation over with your senior commanders. The general consensus of opinion is that you should fall back towards Mersa Matruh. You know the terrain extremely well and you know that soft sand to the south will protect your flank. The only other option is to continue to try to hold further to the east, but your new front would be perilously exposed to a flanking move by the enemy. Your commanders have mixed opinions and ultimately the decision has to be yours. If you want to remain in position then go to **Decision 394**. If you think that you need to pull back toward Mersa Matruh then go to **Decision 264**.

352

Your predominantly-German force falls back towards Baggush. Most of the Italian troops have now either been killed or have been forced to surrender, having been cut off or overwhelmed. What remains of your army is a very sorry force. Now turn to **Decision 332**.

353

In deciding to wait the enemy has taken the opportunity to seize Baggush, to your rear in the west. However, your intelligence reports suggest that it is only thinly held and that there is yet an opportunity to break through if you are quick. If you wish to take this opportunity then go to **Decision 397**. If you still feel that you need Hitler's blessing to fall back then go to **Decision 396**.

354

You touch down on a desert airfield on the evening of 25 October. You are immediately bombarded with information and you also hear of the untimely death of Georg Stumme. He has apparently been travelling in a radio car near to the front on the previous day, heading for a command post. His car came under attack and although Stumme was not injured he suffered a heart attack and died. His body was not found until this morning. Temporarily von Thoma has assumed command of the army.

You make for your command vehicle, a captured British Dorchester truck. There have been enormous casualties in the north due to extensive enemy artillery and aircraft attacks. The Italian Trento Division has been badly mauled, the German 164th Division has lost two battalions, the 15th PZ and the Littorio Division have lost several of their tanks. All of your units in the north are under strength and what is more you have only sufficient fuel for three days of operations. The crucial Point 29, an observation post at the edge of a minefield, has been taken by the enemy. It was at the heart of the 164th Division's perimeter.

You now need to make decisions very quickly. If you wish to counterattack Point 29 now then go to **Decision 375.** If you want to wait and assess the situation more fully then go to **Decision 292.**

355

2000 hours 26 October 1942

Having decided to continue with the counterattack against Point 29, it is only a matter of hours before you realise just how overwhelming the enemy forces in the area have become. Each attack by your troops is quickly countered by increasing numbers of enemy tanks. By late evening it has become clear that the 15th PZ has virtually collapsed as a fighting unit and that the enemy is about to break through. This will mean that what is left of the 164th Division and the 7th Bersaglieri, protecting the coast, will be cut off. With the enemy being so aggressive, you have little time to make up your mind. If you want to order the 164th and the Italians to hold then go to **Decision 268**. If you are prepared to authorise their withdrawal towards Sidi Abd el Rahman then go to **Decision 293**.

356

0900 hours 27 October 1942

You order them to hold their positions around Sidi Abd el Rahman. The new day brings alarming reports that the enemy has pushed a second armoured division into the coastal strip area. Your troops are continuing to hold and are trying to establish a cohesive line. However from the reports being passed from the battalions to the divisions the situation is becoming critical. Several battalion commanders are requesting support, claiming that it is impossible for them to establish a new line. If you want to order the 90th Light Division to move east to reinforce the coastal strip then go to **Decision 336**. If you want to give the 164th and the Italians permission to fall back towards Ghazal then go to **Decision 315**.

357

2400 hours 27 October 1942

In ordering a withdrawal you take the precautionary measure of ordering the 21st Panzer and the Ariete Division to fall back towards Fuka to establish a blocking position. Your other units are instructed to begin their withdrawal overnight. You fear for the Italian infantry divisions; they lack any transport and any retreat will have to be undertaken on foot, in the darkness. If they have not reached safe positions by dawn they will be hopelessly exposed to enemy air attacks. As the night draws in and you monitor radio traffic you hear that the enemy has raced columns of tanks and infantry and has seized Ghazal and El Daba. If you want to re-route the 21st PZ Division to help cover the retreat then go to **Decision 270**. If you think that this is a forlorn hope and will simply fritter the unit away then go to **Decision 245**.

358

You decide to order the retreat now, before it is too late. You re-establish your command centre and almost immediately receive a coded message from Hitler himself. He is absolutely livid with you and blames you for having lost the army in North Africa. He tells you that on every other front the German forces are victorious and advancing. You are to hand over your command to General von Thoma with immediate effect and you are to present yourself to Hitler in Berlin. Undoubtedly this is the end of your military career, but your fate may actually be worse.

359

2400 hours 28 October 1942

The two armoured units move into position, covering the Rahman Track. They are barely given time to dig in and prepare their positions before the enemy begins probing the area with armour and plastering troop concentrations with aircraft attacks. You now realise that the south is in grave danger. The enemy is making continued progress along the coastal strip. If they are able to take some of the coastal towns then no matter how determined your defence is further south, all of your army will be cut off. With great reluctance you decide to fall back to try to re-establish a firmer front line to the west. Now turn to **Decision 248**.

360

The Littorio Division is under huge pressure in this sector, but they are holding. You decide to move up the 21st Panzer Division and the refitted 15th Panzer Division. They manage to replace the Littorio Division to hold the line protecting the Rahman Track. Now turn to **Decision 382**.

361

This is not working. You will need to commit your mobile forces to cover the gap from the Qattara Depression in the far south to Jebel Kalakh, otherwise the enemy will simply flank your entire line. If you wish to move the Ariete Division into position then go to **Decision 276**. If you want to move the 21st Panzer Division into position then go to **Decision 251**.

362

The situation in the north is deteriorating. The enemy has pushed armoured columns and Lorried Infantry along the coast road and has seized Ghazal and El Daba. Your troops are desperately trying to establish a new front at Fuka. To the south, Italian infantry divisions are slowly moving northwest onto the Telegraph Track, but are now forced to cross the desert itself, due west, to avoid being cut off. Many of the Italian units are hopelessly exposed and easy prey for the marauding mobile enemy formations. You could risk sending the 21st PZ Division to help them, in the hope that more of the infantry can be saved. If you want to send the 21st Panzer to counterattack and do what they can, then go to **Decision 270**. If you want to order the 21st Panzer to try to escape to the west and retain it as an effective fighting unit then go to **Decision 245**.

363

You decide to launch an assault against the salient, but you need to decide whether to attack the Kidney Ridge feature or focus on the northwest or the southwest of the salient. You cannot hope to launch a general counter offensive. By focusing on one area you can hope to gain temporary superiority in terms of armour. If you want to launch your attack against the Kidney Ridge feature then go to **Decision 277**. If you want to attack the northwest of the enemy salient then go to **Decision 300**. If you want to attack the southwest of the salient then go to **Decision 321**.

364

Your solid and determined anti-tank gun screen successfully beats off several attempts by the British armour to break through. As the enemy tanks withdraw your anti-tank line comes under heavy artillery bombardment and is buzzed by flights of enemy aircraft. The enemy aircraft are shooting up transports and any exposed positions. Suddenly the front seems to quieten after the ferocity of the tank attack, artillery bombardment, and aircraft attention. If you want to continue to hold your positions and await developments go to **Decision 279**. If you want to launch the 90th Light Division against Point 29, held by Australian troops, then go to **Decision 322**.

365

2400 hours 29 October 1942

The Australians continue to probe forward around the Point 29 positions. There are rumours that Montgomery is pulling units back into a reserve. If this is true he is clearly preparing to launch a major assault somewhere along the front. This will almost certainly fall in the northern sector. If you think Montgomery is aiming to focus on Point 29 then go to **Decision 303**. If you think he is more likely to launch a renewed offensive from the Kidney Ridge salient then go to **Decision 281**.

366

In the darkness, somehow, the 21st Panzer Division managed to punch through the relatively-thin Australian defensive line and break through to the salient. Von Thoma tells you that his commanders on the ground anticipate that the enemy will launch aircraft attacks at daybreak and possibly support these with armoured thrusts. He is of the opinion that the 21st Panzer Division will not be able to keep the route through to the salient open for very long in the morning. They are already under pressure and expect this to intensify. You need to decide whether the troops in the salient should remain in position or not. If you want to take the opportunity to evacuate as many of the men as possible and try to re-establish a line at Sidi Abd el Rahman then go to **Decision 257**. If you want to order the 21st Panzer to keep the route open and for the men in the salient to hold their positions then go to **Decision 282**.

367

2400 hours 1 November 1942

The situation in the south is extremely confused; understandably so, as all of your German and Italian units are on the move, desperately fighting rearguard actions as they withdraw towards the northwest. The enemy now seems to be advancing in force along the whole of the front. Scattered armoured units and elements of the 90th Light Infantry Division are engaged in a major fight for Ghazal. They have had barely enough time to take up positions before the lead elements of the enemy armoured divisions fall on them. At the moment the situation seems to be quite promising; the new line at Ghazal is holding. But you fear that this cannot last for long, particularly when daylight comes and the enemy is able to redeploy its artillery further west and then launch air attacks against the positions. It seems to you that you have three choices. You can continue to try to hold out at Ghazal, giving your troops in the south the maximum opportunity to withdraw to relative safety, by going to **Decision 258**. You could order a withdrawal now towards Fuka, but this would mean that the bulk of your forces in the south would be lost. If you wish to do this, go to **Decision 283**. If you wish to consider your own position and resign your command of the Afrika Korps, having mismanaged the battle, you should go to **Decision 379**.

368

0100 hours 2 November 1942

The enemy launches a huge attack to the northwest of the Kidney Ridge salient. There is a seven hour aerial bombardment of Tel el Aqqaqir and Sidi Abd el Rahman. Large swathes of your front are shattered by pinpoint aircraft attacks and your supply lines are disrupted by accurate enemy artillery fire. What remains of your front line force brace themselves for the inevitable assault by armour and infantry. Where did you concentrate your armour? If you concentrated them on the coastal strip then go to **Decision 390**. If you concentrated them to face the Kidney Ridge salient then go to **Decision 259**.

369

1100 hours 2 November 1942

Just as von Thoma had predicted, your armour begins running out of fuel before it even reaches its new positions. The Italian armour has been particularly affected, as their fuel supplies were even more critical than those of the German forces. The enemy meanwhile has broken through your front line and is pouring forces towards Tel el Aqqaqir. All now rests on your positioning of your armour and whether or not you should commit to a counterattack. You know that in doing this you run the risk of losing it all should it run out of fuel. If you want to commit your armour then go to **Decision 286**. If you want to call off the armoured engagement then go to **Decision 328**.

370

You order your forces to counterattack. To begin with they make fair progress, but the enemy is feeding in more armoured units and eventually your counterattack loses all momentum. General von Thoma reports that the armour is running out of fuel and that increasing numbers of tanks are having to be abandoned on the battlefield. Suddenly you lose radio contact with him. Now you feel blind and unable to gauge how the battle is progressing. Your staff try to establish contact. They tell you that your forward positions have been overwhelmed and that the enemy is advancing west and northwest, with very little between them and your vulnerable rear areas and supply routes. It has been a valiant effort, but it has failed. You order a general retreat to the northwest, in the certain knowledge that the bulk of your army has been lost. You receive communication direct from Berlin, ordering you to make your way to the airfield at Mersa Matruh, where a Ju52 will be waiting to evacuate you and your command staff. You are to fly to Naples and from there to Berlin and present yourself to Hitler. He will not countenance the capture of a German Field Marshal. You know that your military career has come to an end.

371

Your armour has just sufficient fuel to extricate itself. Slowly but surely they move west. General von Thoma tells you that your effective armoured force is now down to around forty vehicles. Even these are low on fuel and the crews report not having seen any fuel bowsers for nearly twenty-four hours. You know only too well that many of them lie alongside the tracks and roads, having been wrecked by enemy aircraft attacks. You now need to make the decision as to whether or not you will try to hold the line where you are, in order to try to re-establish some kind of stability along the front. If you want to try to re-establish the line, then turn to **Decision 349**. If you think a withdrawal is necessary before all is lost then go to **Decision 285**.

372

You decide to withdraw, but the critical question is whether you will choose to pull back the entire army and go to **Decision 393**, or whether you will focus on trying to extricate the German contingent and go to **Decision 263**.

373

You fall back towards Baggush. Your decision to do so is only just in time, as the enemy launches a major attack on Fuka, overwhelming your rearguard. The enemy barely pauses and presses on from the town, snapping at your heels. There is no opportunity for you to turn and fight. You need to keep going, otherwise you will be encircled. You temporarily hold at Baggush, but know that to make any further retreat will need the personal intervention of Hitler. If you want to try to hold Baggush for as long as possible and wait for permission to fall back on Mersa Matruh then go to **Decision 374**. If you feel you have no option other than to withdraw to Mersa Matruh now then go to **Decision 290**.

374

You decide to hold at Baggush, but the enemy is closing in and you know that the end of your army is close. You can either decide to try to fight your way out with what limited resources you have left and go to **Decision 400**, or you can put an end to the unequal fight and drive forward and offer your surrender to Montgomery by going to **Decision 398**.

375

1500 hours 26 October 1942

You pull together elements of the 15th PZ, 164th Infantry and parts of the Italian XX Corps to launch the counterattack. Even while your troops are assembling they are brought under tremendous artillery fire and incessant air attacks. Your men have not even left the start line and casualties are already mounting. If you wish to go ahead with the counterattack then turn to **Decision 242**. If you want to call off the counterattack then go to **Decision 267**.

376

2000 hours 26 October 1942

You decide to call off the counterattack, realising that you will simply fritter away your most able and mobile troops in a war of attrition that you cannot possibly win. The 15th PZ has taken enormous casualties; large numbers of their vehicles are not operational. However in calling off the counterattack your engineers have been able to extricate many of the damaged and broken-down vehicles. In time they can be repaired. What this does mean, however, is that the 15th PZ is unavailable to you for at least forty-eight hours and even then it will only be a shadow of its former self. Now turn to **Decision 243**.

377

0900 hours 27 October 1942

You decide to reinforce the front with the 90th Light Division. By morning it does appear that the 90th has succeeded in stabilising the new front line. They report that the line is under considerable pressure from increasing numbers of enemy tanks and Lorried Infantry. As you look through the reports and study the map it is clear that the situation is very dangerous. You cannot afford to allow the enemy to gain control of the coastal road; it will cut off all of your troops, particularly those in the south. It may be time to consider a general withdrawal. If you want to order your troops in the south to begin a withdrawal before the situation becomes critical then go to **Decision 296**. If you think that the front is not as imperilled as it might first appear and you do not want to order a general withdrawal then go to **Decision 317**.

378

2400 hours 27 October 1942

Earlier in the evening the enemy launched a massive bombardment to the west of the Kidney Ridge area. They then pushed an armoured division, backed by an infantry division, through the gap in your defensive line. Effectively your troops in the south are now cut off. You need to decide what to do about the situation, as it may mean sacrificing the last of your mobile units to stem the tide. If you want to order the 21st Panzer Division and the Ariete Division to launch a counterattack against the lead enemy units, which are now believed to be streaming west and north, go to **Decision 271**. If you want to order the two divisions to escape and try to form a blocking position further west then go to **Decision 246**.

379

With huge regret you draft your resignation and order it to be sent to High Command, with a copy to Hitler himself. You feel that you have no other option. Had you been supplied and had replacements and reinforcements been sent then the battle of El Alamein may have resulted in an entirely different outcome. For months now you have realised that North Africa, as far as Hitler is concerned, is a backwater. He has been obsessed with the Russian campaign and has thrown all of Germany's resources into that conflict. The Italian troops, whilst having fought bravely, have also been poorly supported by their own command structure. You hand over command to General von Thoma and await your fate.

380

Ignoring Hitler's demands for your resignation you lead a tattered remnant of your army back to Mersa Matruh. The enemy has not closed with you; he has over-extended his supply lines and you know that you have a short period of time to try to create a defensive position. You are not given the time, however. You receive an unexpected visit from Berlin; two senior army officers present themselves at your headquarters. They have instructions from Hitler to escort you to Berlin, under arrest. Your career is over, as is your command of the Afrika Korps.

381

1200 hours 28 October 1942

The enemy has punched a hole in your defensive line all the way to the Rahman Track. Reports suggest that armoured columns are moving northeast along the track towards Sidi Abd el Rahman itself. The town has been flanked. More enemy armour is pushing towards the Telegraph Track. It is believed to be the 7th Armoured Division. The only effective mobile troops that can respond are the 21st Panzers and the Ariete Division. It is now a question of whether you want to sacrifice them or order them to fall back to try to establish a new line further west. If you want to order them to counterattack and hold open the line of retreat then go to **Decision 271**. If you want to order them to fall back to the west then go to **Decision 246**.

382

The 21st Panzer holds the area of the front to the south of the Kidney Ridge feature. The 15th Panzer is in the centre and north, with the 90th Light Division beyond them in the north. The enemy is clearly trying to establish itself in some force in and around the Kidney Ridge feature. If you want to hold your positions and react to the enemy then go to **Decision 384**. If you want to counterattack against the Kidney Ridge area now and pinch the enemy out of this salient then go to **Decision 252**.

383

2200 hours 27 October 1942

You decide to wait for darkness in order to avoid unnecessary casualties. Your armoured units engage the 7th Armoured Division and manage to hold them. Just as you are about to celebrate you receive intelligence reports that the Free French and Greeks have pushed raiding columns deep behind your armour and have cut off their supply lines. In the darkness it will be almost impossible to track them down and eliminate them. If you want to order your armour to pull back then go to **Decision 275**. If you think that your armour should continue to attack and that you can sort the situation out after dawn then go to **Decision 250**.

384

1200 hours 27 October 1942

Your troops are in relatively secure positions and have dug in to protect the Rahman Track. It is clear that the enemy is establishing its positions around the Kidney Ridge feature. Undoubtedly they will make attempts to enlarge this salient and threaten the Rahman Track and your troops along the coastal strip. It is a question of whether to wait for developments or to take the initiative. If you wish to continue to hold your positions for now and await developments then go to **Decision 342**. If you want to launch an attack against the Kidney Ridge salient then go to **Decision 363**.

385

You decide to call off the assault on the southwest of the salient. Your armour begins to fall back; the losses have so far been around thirty tanks. These are losses that you can ill afford. The enemy is clearly prepared to protect the salient and he is in too great a force for you to dislodge him. You need to shift the focus of your attention and perhaps strike elsewhere. Point 29, slightly to the north, has been taken by Australians. If you want to launch an assault against Point 29 then go to **Decision 322**. If you think it more prudent to remain on the defensive to enable you to respond to developing threats then go to **Decision 278**.

386

You manage to hold off the enemy armour with your anti-tank guns. You are suspicious that the enemy has not pressed home their attacks in this sector with great vigour. This must mean that the enemy is planning something else. If you want to continue to hold and wait for their next move then go to **Decision 279**. If you want to launch the 90th Light Division against the Australians holding Point 29, to the north, then go to **Decision 322**.

387

2400 hours 27 October 1942

Your timely reinforcement has pushed the enemy back in considerable disorder into the original minefield area. The enemy is still launching localised counterattacks along the coastal strip. Your engineers are hard at work salvaging abandoned enemy tanks with the aim of using them as replacements for your losses. You are delighted that enormous amounts of fuel have also been captured. You have completely regained the initiative and despite the enormous numbers of enemy troops they are only putting up a token resistance and continuing to fall back toward their own start lines. You hear a rumour via Berlin that Montgomery has been replaced and that Horrocks has been given command of the 8th Army. You fire off a report to Berlin and receive congratulations on your victory at El Alamein. You determine to return to Berlin and ask Hitler in person to reinforce you. Perhaps there is still an opportunity for the Afrika Korps to burst into Egypt and seize the Suez Canal. Against all the odds you have defeated a considerably larger and more potent army.

388

The combined, largely-infantry attack is disastrous. The 90th Light Division has suffered huge casualties from pinpoint enemy artillery bombardment. It is clear that the enemy is directing fire from Point 29, where they have a clear view of the battlefield. There have been reports of continued use of star shells, which have illuminated your assembly positions. The 164th Division and the Bersaglieri have also suffered, not only in their contribution to the counterattack, but also in attacks from the east where they have been forced to give more ground overnight. The position along the coastal strip is now extremely perilous. Now turn to **Decision 304**.

389

2400 hours 1 November 1942

What remains of your army struggles back towards Fuka, abandoning the coastal strip just ahead of the advancing enemy units. It is readily apparent that Fuka is no safer a position to try to re-establish the line. But the decision to relinquish so much territory to the enemy is not entirely yours. You could order a general retreat towards Mersa Matruh and go to **Decision 247**. With the enemy already threatening Baggush, you could consider trying to hold, but you know that your time as commander of the Afrika Korps is about to come to an end. If you wish to hold go to **Decision 347**.

390

Your armour, positioned in the coastal strip, is taking a pounding from the artillery and aircraft attacks. Fuel dumps and ammunition supply centres have been destroyed and trucks moving up men, equipment and supplies are now burning along the supply routes. Your commanders are reporting increasing losses amongst your precious tank reserves. If you wish to leave them in position then go to **Decision 284**. If you want to move them out of the danger zone to the south then go to **Decision 306**.

391

In the swirling heat and dust of the battlefield your tanks manage to temporarily halt the advance of the enemy armour. Von Thoma reports that he has barely thirty-five running tanks left. The anti-tank gun screen is down to around a third of its original strength. He tells you in no uncertain terms that one more determined attack by the enemy will see the line irrevocably breached. He wants to know what your instructions are and whether or not he is to continue to hold or whether you will countenance a withdrawal. If you want von Thoma to continue to hold with what remains of the armour and the anti-tank gun units then go to **Decision 260**. If you want to give him permission to extricate what he can save then go to **Decision 285**.

392

Having taken the decision to withdraw, the next question is how far to fall back along the coast. You need to ensure that whatever position you take up provides you with a viable defensive position. The town of El Daba lies on the railway track near the coastal road. It is near the end of the Telegraph Track, along which your troops from the south have already begun their retreat. The Telegraph Track could be cut very easily by enemy armoured probes and the position is also vulnerable to a flanking move to the south. If you wish to withdraw to El Daba then go to **Decision 262**. If you feel that El Daba is too close and that you will have insufficient time to prepare your defensive positions before the enemy close in on you then the next option is Fuka, which is further to the west. This will, however, mean that significant numbers of your troops retreating from the southern part of your old defensive line will find it almost impossible to link up with you. If you wish to choose Fuka as your new defensive position then go to **Decision 287**.

393

Having decided to withdraw you have to cope with the reality that the bulk of your troops stationed in the south have been cut off. Elements of your armoured force, along with the 90th Light Division, are engaged in a running battle with lead enemy formations around Ghazal. General von Thoma tells you that for now the troops have an opportunity to hold the line along the coastal strip. But he does not expect to be able to resist for very much longer. If you want to continue to fight on, go to **Decision 288**. If you feel that you have lost all hope of saving your army and wish to tender your resignation as commander of the Afrika Korps, then go to **Decision 379**.

394

Your scattered outposts covering the southern flank report that the enemy is moving extensive columns of armour, Lorried Infantry and supply trucks to the south of your positions. It seems clear that they are aiming to seize Fuka and cut you off. You know that there is little that you can do to prevent them. If you wish to launch limited counterattacks to try to stop them then go to **Decision 311**. If you think that the enemy move gives you no option other than to fall back while you still have a chance then go to **Decision 264**.

395

0800 hours 4 November 1942

It is clear that the enemy is moving in force towards Fuka, with Lorried Infantry and armour. Further columns are stretched out in the desert, aiming to flank you. The situation is dire and encirclement threatens. If you want to continue to wait for permission to withdraw then go to **Decision 265**. If you want to withdraw now then go to **Decision 290**.

396

The enemy closes in on you and surrounds what remains of your command. Units that have fought so well for you on numerous battlefields in the desert are crumbling and surrendering to the enemy. You receive orders from Hitler instructing you to fly west by whatever means. An aircraft will be on standby to convey you to Berlin via Italy. He does not wish you to fall into enemy hands. If you wish to accede to his instructions then go to **Decision 399**. If, however, you think that honour is more important than his directive and that you wish to surrender your army and yourself to Montgomery then go to **Decision 398**.

397

Your ragtag army manages to break through to Mersa Matruh. Your troops are woefully short on equipment, fuel, food and water. You can only hope for a miracle as the enemy formations begin to close in on your positions. Suddenly the sky darkens and a torrential rainstorm begins. The rain continues for hours; this is exactly what you needed. The soft sand to the south, your unprotected flank, is now impassable. You are able to contact High Command and appraise them of the situation. Reluctantly they allow you to begin a phased withdrawal towards El Agheila, where you can establish a better defensive position. The rains have taken the momentum of the enemy offensive from them and you are able to continue to hold for several more weeks. Eventually you are given permission to withdraw into Tunisia, leaving Libya forever in February 1943. You leave with huge regrets, however. Had you been supported then you would have been able to have conquered Egypt, seized the Suez Canal and who knows, you could have had the opportunity to link up with German forces sweeping down from the north. You could have captured the Middle East and brought this war to an end.

398

Taking only one aide with you and flying a white flag on your staff car you expend your last can of fuel heading out beyond your front line, seeking to contact the enemy. You present yourself to a scruffy infantry officer. Within an hour a ceasefire has been negotiated. An hour later you see a British armoured car, flanked by several tanks, moving towards your position in no-man's land. Montgomery has arrived and the surrender negotiations can begin.

399

With enormous reluctance you decide to accede to Hitler's demands that you fly out of Libya and abandon your men to their fate. General von Thoma is to take over temporary command of the Afrika Korps. He will need a miracle to extricate the army from the perilous situation. As for you, you know that your career has come to an end. At best you can hope for house arrest and relative obscurity, at worst a public trial for dereliction of duty.

400

The end of your war in North Africa was, perhaps, already written when you first set foot on the continent; dying at the head of your own troops, in a desperate counterattack against insuperable odds, against an enemy that you have chased across the length and breadth of the continent. Only now, with the odds stacked so heavily in their favour, have they been able to defeat the Afrika Korps. None will know your true fate; just another armoured car burned out in the desert with bodies left inside.